DESIGN AND APPLICATIONS IN DIESEL ENGINEERING

ELLIS HORWOOD SERIES IN MECHANICAL ENGINEERING

STRENGTH OF MATERIALS
Vol. 1: Fundamentals — Vol. 2: Applications
J. M. ALEXANDER, University College, Swansea
TECHNOLOGY OF ENGINEERING MANUFACTURE
J. M. ALEXANDER, University College, Swansea, G. W. ROWE, Birmingham University and
R. C. BREWER
VIBRATION ANALYSIS AND CONTROL SYSTEM DYNAMICS
C. BEARDS, Imperial College of Science and Technology
STRUCTURAL VIBRATION ANALYSIS
C. BEARDS, Imperial College of Science and Technology
COMPUTER AIDED DESIGN AND MANUFACTURE 2nd Edition
C. B. BESANT, Imperial College of Science and Technology
BASIC LUBRICATION THEORY 3rd Edition
A. CAMERON, Imperial College of Science and Technology
SOUND AND SOURCES OF SOUND
A. P. DOWLING and J. E. FFOWCS-WILLIAMS, University of Cambridge
MECHANICAL FOUNDATIONS OF ENGINEERING SCIENCE
H. G. EDMUNDS, University of Exeter
ADVANCED MECHANICS OF MATERIALS 2nd Edition
Sir HUGH FORD, F.R.S., Imperial College of Science and Technology, and
J. M. ALEXANDER, University College of Swansea.
MECHANICAL FOUNDATIONS OF ENGINEERING SCIENCE
H. G. EDMUNDS, Professor of Engineering Science, University of Exeter
ELASTICITY AND PLASTICITY IN ENGINEERING
Sir HUGH FORD, F.R.S. and R. T. FENNER, Imperial College of Science and Technology
DIESEL ENGINEERING PRINCIPLES AND PERFORMANCE
S. D. HADDAD, Associate Professor, Western Michigan University, USA, and Director of
HTCS Co., UK, and N. WATSON, Reader in Mechanical Engineering, Imperial College of
Science and Technology, University of London
DIESEL ENGINEERING DESIGN AND APPLICATIONS
S. D. HADDAD, Associate Professor, Western Michigan University, USA, and Director of
HTCS Co., UK. and N. WATSON, Reader in Mechanical Engineering, Imperial College of
Science and Technology, University of London
TECHNIQUES OF FINITE ELEMENTS
BRUCE M. IRONS, University of Calgary, and S. AHMAD, Bangladesh University, Dacca
FINITE ELEMENT PRIMER
BRUCE IRONS and N. SHRIVE, University of Calgary
CONTROL OF FLUID POWER: ANALYSIS AND DESIGN 2nd (Revised) Edition
D. McCLOY, Ulster Polytechnic, N. Ireland and H. R. MARTIN, University of Waterloo,
Ontario, Canada
UNSTEADY FLUID FLOW
R. PARKER, University College, Swansea
DYNAMICS OF MECHANICAL SYSTEMS 2nd Edition
J. M. PRENTIS, University of Cambridge
ENERGY METHODS IN VIBRATION ANALYSIS
T. H. RICHARDS, University of Aston Birmingham
ENERGY METHODS IN STRESS ANALYSIS: With Intro. to Finite Element Techniques
T. H. RICHARDS, University of Aston in Birmingham
COMPUTATIONAL METHODS IN STRUCTURAL AND CONTINUUM MECHANICS
C. T. F. ROSS, Portsmouth Polytechnic
FINITE ELEMENT PROGRAMS FOR AXISYMMETRIC PROBLEMS IN ENGINEERING
C. T. F. ROSS, Portsmouth Polytechnic
ENGINEERING DESIGN FOR PERFORMANCE
K. SHERWIN, Liverpool University
ROBOTS AND TELECHIRS
M. W. THRING, Queen Mary College, University of London

DESIGN AND APPLICATIONS IN DIESEL ENGINEERING

Editor-in-Chief:

S. D. HADDAD, B.Sc.(Eng.), Ph.D.
Associate Professor
Western Michigan University, USA
and Director, Haddad Technical Consultancy Services
Loughborough, UK

Associate Editor:

N. WATSON, B.Sc.(Eng.), Ph.D.
Reader in Mechanical Engineering
Imperial College of Science and Technology
University of London

ELLIS HORWOOD LIMITED
Publishers · Chichester

Halsted Press: a division of
JOHN WILEY & SONS
New York · Chichester · Brisbane · Toronto

First published in 1984 by
ELLIS HORWOOD LIMITED
Market Cross House, Cooper Street, Chichester, West Sussex, PO19 1EB, England

The publisher's colophon is reproduced from James Gillison's drawing of the ancient Market Cross, Chichester.

Distributors:

Australia, New Zealand, South-east Asia:
Jacaranda-Wiley Ltd., Jacaranda Press,
JOHN WILEY & SONS INC.,
G.P.O. Box 859, Brisbane, Queensland 40001, Australia

Canada:
JOHN WILEY & SONS CANADA LIMITED
22 Worcester Road, Rexdale, Ontario, Canada.

Europe, Africa:
JOHN WILEY & SONS LIMITED
Baffins Lane, Chichester, West Sussex, England.

North and South America and the rest of the world:
Halsted Press: a division of
JOHN WILEY & SONS
605 Third Avenue, New York, N.Y. 10016, U.S.A.

©1984 S.D. Haddad and N. Watson/Ellis Horwood Limited

British Library Cataloguing in Publication Data
Design and applications in diesel engineering. –
(Ellis Horwood Series in mechanical engineering)
1. Diesel motor – Design
I. Haddad, S.D.   II. Watson, N.
621.43'62   TJ795

Library of Congress Card No. 84-4576

ISBN 0-85312-733-6 (Ellis Horwood Limited)
ISBN 0-470-20074-X (Halsted Press)

Typeset by Ellis Horwood Limited.
Printed in Great Britain by The Camelot Press, Southampton.

Table of Contents

Chapter 3   The Passenger Car Diesel Engine – Present and Future
W. M. Scott, Ricardo Consulting Engineers plc

Chapter 4   Deformation and Stress Analysis of Engine Components using Models
H. Fessler, Professor of Experimental Stress Analysis, University of Nottingham

Chapter 5 Practical Applications of Finite Elements in the Stressing of Diesel Engine Components
       A. K. Haddock, Perkins Engines

8   Contents

Chapter 6   Crankshaft Loading and Bearing Performance Analysis
               B. Law, Perkins Engines

Contents

9

Preface

When preparing an appropriate introduction to this book, I thought 'Why not deviate from the norm and relate my own growing association with the diesel engine?'. After all, a life cycle such as mine is very much analogous to that of the diesel cycle.

Compression and accumulation of knowledge starting from childhood days when my father took me to visit his remote village (Mar Yaco) in the midst of the rugged Northern mountains of Mesopotamia. I recall visiting the village black-smith where I noticed some old engines and gadgets. One small engine apparently awaiting repair attracted my attention. My father called it a diesel but he said that farmers in that part of the country had been known to produce their own innovative engines, sometimes made from hard wood as well as metal. This simple conversation was imprinted in my memory till it started to make more sense when I later studied the history of the Land of the Two Rivers which is the cradle of civilization. Further compression of knowledge was to follow from completing my university studies in mechanical engineering in England and later from my considerable training and engineering experience with some bias towards engine maintenance and applications at Daura Oil Refinery in Iraq.

Ignition. Attendance at a special intensive four-month course on diesel engineering in 1968 at the UNIDO Oil Engine Research Insitute, VUNM, Prague, instructed me and further enlarged my experience. Subsequently, I was fortunate to work with one of the leading diesel advocates, Professor Theo Priede, at the Institute of Sound and Vibration Research in Southampton University. Both VUNM and Priede ignited my latent love for engines and the diesel type in particular.

Combustion. My awareness of this aspect grew because I continued to develop diesel knowledge at ISVR Southampton, Lucas CAV and Loughborough University, and through my own consulting activities with Haddad Technical Consultancy Services. Of course, many catalysts served to strengthen my commitment but I feel that my wife, Basima, deserves special mention for her unfailing support through thick and thin.

Expansion. In spite of the fact that my associates, my research students and I have, between us, completed much diesel engine research, there remains much to be done. We all hope to continue this expansion process for many years to come.

Acknowledgements. Naturally, a number of establishments and persons have been instrumental in assisting me with this process:

UNICEG: Universities Internal Combustion Engines Group: for providing the opportunity to enhance the propagation of I.C.E. knowledge.

Loughborough University: for providing me with the opportunity to run intensive post-experience courses and facilities for research and teaching expertise.

The Center for Professional Advancement: for providing the opportunity to propagate my courses worldwide.

Dr. Neil Watson: my co-editor for his invaluable contribution towards these two books.

Mr. Ellis Horwood: our publisher for his constant encouragement and advice.

The earlier, companion volume (Principles and Performance in Diesel Engineering) presented a basic introduction to the theoretical and operational background, so initiating the publication of these two complementary books on diesel engineering. In the author's view, it is becoming apparent that yet a third volume is needed to include other advanced aspects of this branch of engine technology.

And finally, on the eve of my departure to the United States, I look forward to further developments at Western Michigan University, and in the diesel engine industry in the U.S.A.

As for *Exhaust,* I would rather like to believe this will not happen; but a life cycle is just like a diesel cycle: it has to come to an end some day, allowing for another cycle to begin . . . and so on. Perhaps my exhaust stroke should be left to the imagination and speculation of the reader.

S. D. Haddad

1

Some basics of high speed diesel design

Martin H. Howarth, Atlantic Research Associates

This is a very wide title and could only be properly developed in a complete book rather than a single chapter. However, many aspects of diesel engine design are covered by other chapters in this volume such as those on structure and noise, lubrication and wear, exhaust emissions, stress analysis, bearing loadings and fuel/air mixing. This chapter therefore deals with the following design concepts:

1. Engine Cycles.
2. Combustion Systems.
3. Thermal Stresses.
4. Motoring Losses.
5. Overall Design.
6. Timing Drives.
7. Cylinder Head and Gasket.
8. What's New?

1. ENGINE CYCLES

Although two-stroke high speed diesel engines are manufactured and although they offer a greater potential of power/weight ratio and power/space ratio than four-stroke engines, their various problems have meant that they are scarcely considered in the small engine field and little development work is in progress. Therefore only four-stroke engines are considered here. The engines as we know them today have been under development for more than 50 years but are still subject to continuing refinement, and new developments are necessary to meet more exacting requirements as they arise.

 The four-stroke diesel engine lends itself extremely well to supercharging and this offers great advantages to the manufacturer by extending the output range of any one basic engine design. Supercharging is therefore likely to extend down the engine size range even to the smallest sizes, whether it be by means of a turbocharger, a pressure wave supercharger or a newly developed type of positive displacement supercharger. The effect of supercharging on diesel engine design is mostly a matter of minimizing heat flow and thermal stresses and the injection equipment. Fundamental design changes are not involved.

Turbocharging is undoubtedly the most attractive way of increasing engine output and that is the reason that it figures quite largely here. As far as the designer is concerned the effect of turbocharging reflects on:

Manifolding—inlet and exhaust
Heat Flow
Maximum Cylinder Pressures
Lubrication

Apart from the bonus it provides in greater output and greater specific output, a turbocharger is in many instances an excellent silencer.

A more questionable feature is whether turbocharging actually results in lower cost per kilowatt of output. One thinks that it ought to do so but figures published recently (that may not have been representative) did not suggest that the cost of a turbocharged engine compared favourably with the cost of a larger, naturally aspirated engine.

The principle of a pressure wave supercharger is shown in Fig. 1. In addition to the design requirements listed above for the turbocharger, it is also necessary to provide a mechanical drive. Even though the drive has little power to transmit, its provision is a considerable nuisance and is only likely to be regarded as acceptable if the supercharger can provide quite substantial advantages such as improved low speed torque.

Early forms of pressure wave supercharger, far from providing any silencing effect, were appallingly noisy. This has been improved by arranging the gas passages in helical form and not as shown in the figure.

As regards supercharging by mechanically driven pumps, the position is not particularly encouraging. A drive transmitting considerable power must be provided and there is likely to be a drop in overall efficiency of the engine/ supercharger combination. Recently there seemed to be interest among component manufacturers to develop more efficient compressors for use as superchargers and this might change the situation. Unfortunately, that interest seems to have waned.

Other methods of improving the four-stroke cycle have been, or are being, investigated. Of these, the diesel Wankel need not distract us now. The so called 'adiabatic engine' in which the aim is to keep the heat inside the engine by the use of ceramic materials and thus improve thermal efficiency and avoid the need for a cooling system is still in its infancy. Although this may well be a most significant development, generally accepted design principles have not yet been

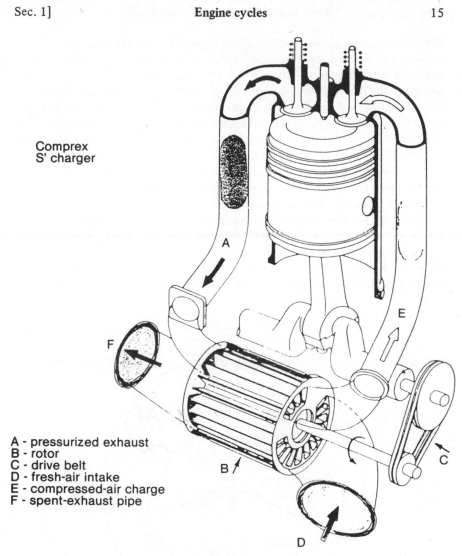

Comprex
S' charger

A - pressurized exhaust
B - rotor
C - drive belt
D - fresh-air intake
E - compressed-air charge
F - spent-exhaust pipe

Fig. 1 – Principle of pressure wave supercharger.

established and it is too early to deal with the subject now. However, much research and development work is currently in progress.

This chapter therefore is confined to four-stroke engines of quite conventional form.

2. COMBUSTION SYSTEMS

Both direct injection (DI) and indirect injection (IDI) combustion systems may properly be considered as having high speed capability. It is well known that the

IDI system has at present greater speed capability and has little difficulty in operating up to the highest rotational speed necessary even for a vehicle. The DI engine at moderate speeds offers some 15% better fuel economy than the IDI and great efforts are being made to extend its speed range to equal that of the IDI without subsequent loss of its good economy and other virtues. The practical speed range of DI engines has been limited to about three to one and this means that the DI engine has not yet made inroads into the passenger car market. The high speed DI engine's problem falls into one of fuel and air matching.

At this stage it is perhaps worth considering the sort of time scale that has to be allowed for injection and the establishment of combustion.

Figure 2 shows a typical timing diagram for a four-stroke engine operating at 3000 rpm. The time available for injection is 1.1 milliseconds and combustion, which has to start before injection is complete, lasts only about 1.4 milliseconds.

At such a speed, satisfactory conditions could be achieved at present by all three current types of direct injection/combustion chamber combination. That is to say, a pump, pipe and injector plus multi-hole nozzle and toridal chamber, or

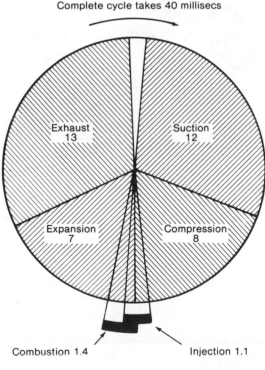

Fig. 2 – Four-stroke diesel engine: cycle of events at 3000 rpm.

a unit injector plus multi-hole nozzle and toroidal chamber, or finally a wall
wetting type of combustion system. For higher speeds up to 5000 rpm, which
system will win out?

It is difficult to see any clear advantage for any of them. A pump injector
system seems seductive at first sight. Figure 3 shows an example; but detailed
consideration reveals a number of disadvantages. The unit injector would make
special demands on the overall engine design and this would reduce the degree of
commonality between diesel and gasoline engines of similar size. For small
engines, that factor would be likely to increase the cost of the diesel version and
thereby diminish its attractiveness. Also, initial assembly and subsequent servicing

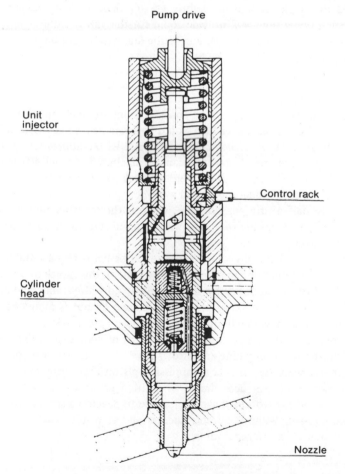

Fig. 3 – Example of pump injector system.

of the injection equipment itself would seem to involve greater costs than those of a separate pump and injector system. Nevertheless, successful engines with unit injectors have been made for many years.

Apart from the very high speed aspect, general design considerations are largely similar for both DI and IDI systems and the two will be treated together here except in so far as either system raises particular problems of detailed design.

Important initial considerations for any engine design are as follows.

2.1 Bore and Stroke
For most systems, approximately 'Square' dimensions are favourable but for preference the stroke should be slightly greater than the bore, say up to 1.15 X bore. Such dimensions will provide room for the valves without undue dead volume above the piston. For a DI engine the diameter of the bowl in the piston should be very close to 50% of the bore.

2.2 Compression ratio
In both DI and IDI engines, practical considerations are likely to ensure that the compression ratio will be higher than would be desirable for maximum efficiency. Although the air cycle efficiency continues to rise with compression ratio, losses due to high pressures rise faster and cause a decline in overall efficiency over about 15:1. In the IDI engine the compression ratio will be determined by cold starting and noise considerations and will be at about 22:1 for engines of approximately half a litre per cylinder (which is the territory mostly occupied by IDI engines) whereas the DI engine with its good starting characteristic might be four ratios lower.

Apart from cold starting which can, in fact, be helped by a suitable starting aid such as a glow plug or a manifold heating system, the choice of compression ratio will be determined partly by the need to avoid white smoke and misfire at part load. This becomes a major problem as engine output is increased because the compression ratio will be limited to a low value in order to keep the maximum cylinder pressures within a reasonable limit. It is, of course, a problem that has been recognized for a long time and efforts have been made to overcome it. One such effort has been the variable compression piston that automatically adjusts the compression ratio in accordance with firing pressures. Very considerable efforts by competent workers have been made to develop a satisfactory variable compression piston without notable success and it would seem that the idea must be written off as over-ambitious.

2.3 Maximum Cylinder Pressures
These will constitute a problem in almost any engine design and particularly so when a diesel engine is arrived at by converting a gasoline engine. IDI engines

generally exhibit lower maximum cylinder pressures than DI engines, in spite of their higher ratios. Swirl chamber engines, as exemplified by the Comet type combustion chamber, ideally have a maximum cylinder pressure about 20% greater than the compression pressure. However, by retarding the injection, maximum pressures little greater than the compression pressure may be maintained with very little loss in combustion efficiency. This is not feasible with DI engines to the same extent.

The effects of choice of compression ratio and injection timing are summarized on Fig. 4.

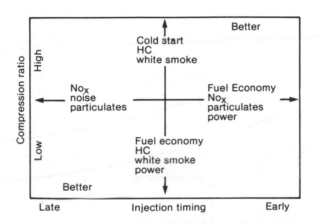

Fig. 4 – Effects of compression ratio and injection timing.

3. THERMAL STRESSES

Components subject to high rates of heat flow and therefore the likelihood of high thermal stresses are obviously the cylinder head, the cylinder liner or barrel and the piston. Pistons will present some degree of problem in virtually all engines, water cooled cylinder heads and barrels are less of a problem in small engines, but those above, say, 120 mm bore will also require great care in design as engine ratings rise. Air cooled cylinder heads and barrels always present a challenge and for a vehicle engine rating meticulous design and extremely high quality foundry work are required.

Figure 5 shows a suggestion for acceptable ratings for pistons, both uncooled and with different forms of cooling related to power per unit area of piston and bore size. This is for light alloy pistons. To form the cast-in gallery which seems the most effective method of cooling such pistons at high rating, the best method would seem to be the use of soluble cores rather than the use of a built-up piston. For higher ratings, a nodular iron piston seems favourable and

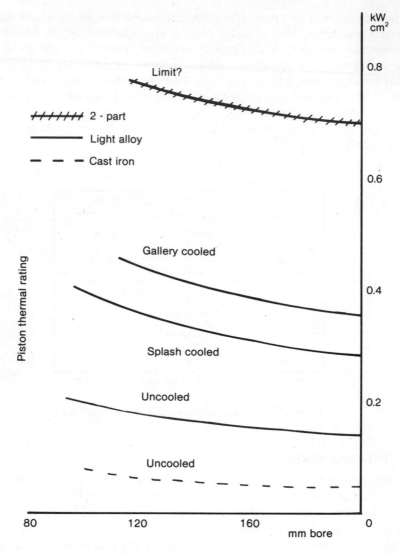

Fig. 5 – Acceptable ratings for pistons, both cooled and uncooled.

for the highest ratings a two-part piston with light alloy skirt and heat resisting crown is the best solution. This involves considerable complication and expense. Fig. 6 shows a gallery cooled piston.

The piston is certainly a multi-function component. Many requirements for it are conflicting. In order to conduct heat from the crown and also for adequate support of the gudgeon pin, generous metal sections are required, but this is in conflict with the need for minimum mass. To keep the height of the engine to

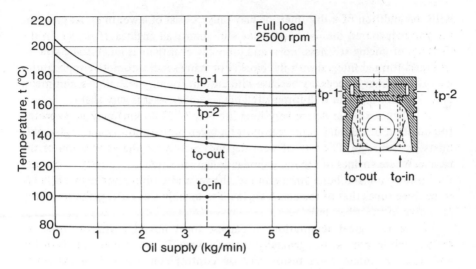

Fig. 6 – Gallery cooled piston.

a minimum the piston should be short, but this may give rise to excessive piston motion within the bore and problems of oil control and piston slap. The piston design is therefore a matter of great importance in the further development of diesel engines. One method of helping to resolve some of these conflicts is by the use of what is effectively a bi-metallic system by incorporating a strip of a different material – an old idea, but one now popular in diesel engines.

Another idea is to reduce the contact area of the skirt as much as possible and in the ultimate this can be reduced to a series of load bearing pads.

The temperature of the top ring is a critical aspect of piston design. A maximum temperature of 260°C is acceptable with a rectangular ring, a few degrees more with a trapezoidal or keystone ring.

The piston must support and locate the gudgeon pin which is normally of the fully floating type. Circlips or snap rings are used and often the grooves for these are formed in the gudgeon pin bore in the piston. This is somewhat wasteful of the available bearing area and a better solution has been shown to be circlips mounted at the end of the gudgeon pin, carried in grooves on the pin itself and bearing against faces machined in the piston. The gudgeon pin diameters range from about 33% of the engine bore, which is normal for naturally aspirated engines, up to as much as 47% of the bore for highly supercharged engines where maximum cylinder pressures approaching 200 bar may be expected at pressure ratios of over 4:1.

3.1 Cylinder Barrel

Heat from the piston must pass through the walls of the cylinder barrel to the coolant. The cylinder barrel may be formed in the cylinder block itself, possibly with the addition of a dry liner, or may only consist of a wet liner. At one time the writer favoured the use of wet liners in almost all engines. However, in the interests of engine stiffness, noise and probably cost there is much to be said for cylinders formed integral with the block. For many small automotive applications no cylinder liner is necessary because with current foundry techniques, chromium plated piston rings and well formulated oils, cylinder wear is very slight.

For larger engines where very long life (say 650 000 km) may be expected the question of cylinder bore wear must be taken more seriously and replaceable liners are desirable. The rate of wear depends greatly on the application of the engine. Where engines of generally similar design are fitted respectively to coaches for highway use and buses for urban use, the wear rate of the bus engine is likely to be three times that of the coach engine. Presumably this reflects the start/stop operation of the urban bus.

There is a great temptation to siamese the bores of engines in order to reduce their length but this generally proves unwise. In order to reduce distortion and hence maintain good piston and oil control conditions, it is extremely desirable to maintain a circulation of water completely round each cylinder.

For very highly rated engines the wet liner remains the best solution. Here the need to contain high cylinder pressures is in conflict with the need to extract heat, and materials of high thermal conductivity are not appropriate. Plain cast-iron liners cannot be used as with a wall thickness sufficient to give satisfactory mechanical stresses the temperature gradient will ensure the liner temperature at the top ring level is excessive. Two possible ways are to provide cooling passages, so that the liner itself is directly cooled by coolant or by oil, or to employ a relatively thin cast-iron liner with a steel reinforcing band outside it.

Cast-iron is a very satisfactory bearing surface for piston rings and for this reason it has survived in general use for a very long time. Efforts in the past to improve the wear qualities by surface treatment or by chromium plating have not in general been thought to be worth while. More recently, the use of silicon carbide applied to the pre- finished liner surface has shown great promise in reducing wear rates under arduous operating conditions and this technique may come into wider use.

3.2 Cylinder Head

The cylinder head presents problems of design, particularly in respect of gasket sealing and of limiting distortion (especially if an overhead camshaft is involved) but even at very high ratings, thermal stresses should not be a limitation providing that well established rules are followed. The rules are to provide a brisk and controlled water circulation all over the cylinder head deck and to ensure a high water velocity through areas of high heat flow, such as between the valves, by providing drilled passages and a carefully planned route. The thickness of metal sections within the cylinder head should be kept to a minimum by good foundry practice and by machining where possible, i.e. by milling back the bridge between the valves.

One advantage of drilled passages for water cooling within the cylinder head is that the coolant is in contact with the head material without suffering from any thermal barrier that might be imposed by the 'as cast' surface of the cylinder head casting. With cast-iron as a cylinder head material the thermal barrier effect can be quite pronounced. Figure 7 shows the temperature gradient through an 'as cast' iron head from the flame side to the coolant side whereas Fig. 8 shows the temperature gradient under the same conditions but with the skin machined away on the coolant side.

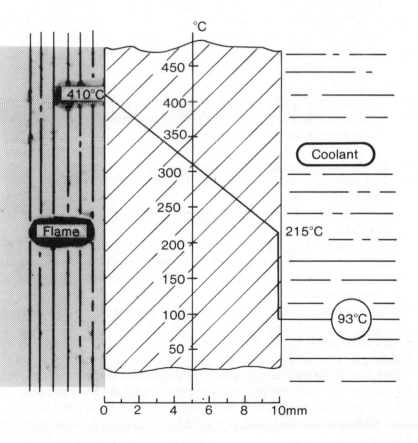

Fig. 7 – Temperature gradient through cast-iron cylinder head deck: coolant side 'as cast'.

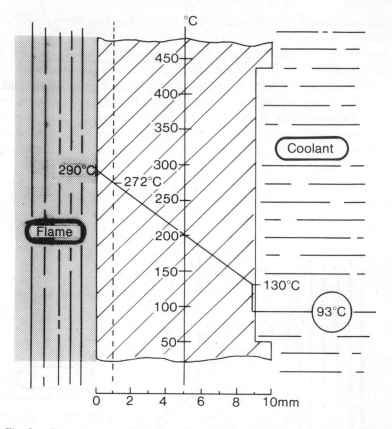

Fig. 8 – Temperature gradient through cast-iron cylinder head deck: coolant side machined.

Figure 9 shows a water cooling jet in a two-valve swirl chamber head. It will be agreed that this is a long jet and it is remarkably difficult to persuade water to flow through such a jet as this with a velocity sufficient to be effective. If the velocity is not adequate the water will boil locally and the resulting steam pocket will further disrupt the heat transfer arrangements such that the situation will be worse than if no jet had been provided because of the great thickness of metal that is required to form the jet.

In order to ensure a sufficient velocity through the jets requires considerable care in sizing all the water passages between the cylinder block and head. If the area of the passage ways is kept very small to try to achieve a high velocity, the pump flow may be throttled and thus the total mass flow of coolant will be inadequate.

Even the most thoughtful design may not result in a satisfactory water circulation system and it may well be necessary to make final adjustments to the

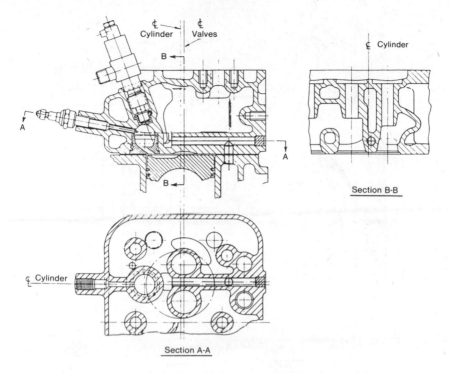

Fig. 9 — Water cooling jet in two-valve swirl chamber head.

sizes of the various passages following measurements of actual flow rates. It is in fact extremely difficult to measure local water velocities within a cylinder head or even to observe currents and stagnant areas. A useful technique is to carry out the flow tests with air instead of water, as velocities may then readily be mapped with pilot heads. To allow for the difference in Reynolds number, the air flow should be about 13 times that of the water flow.

Figure 10 shows cooling jets suitable for a four-valve DI head.

Unlike the case of the cylinder barrel, high conductivity materials can be used for the cylinder head. Light alloy is an excellent material for diesel engine cylinder heads for engines up to at least 120 mm bore. Its use provides a bonus in the form of considerable weight saving as well as superior heat transfer characteristics.

3.3 Modelling

The detailed design of these critical components cannot be dealt with here as, of course, it will be influenced by many considerations resulting from the overall engine design. Nevertheless, the combination of high cylinder pressures and high thermal stresses imposes a limit to specific engine performance, and the limit can only be pushed back by improvements in the design of these components. The

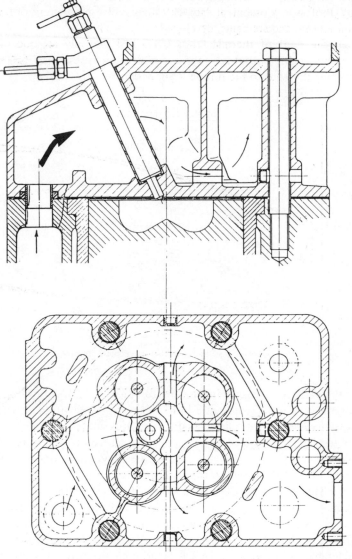

Fig. 10 — Suitable cooling jets for four-valve DI head.

importance of the subject has lead to a great deal of work on computer modelling to determine expected pressures, temperatures and heat flows and it would be unwise to embark on a development programme for any highly rated engine without such aid. Curves showing the close correlation between measured results and those predicted by calculation are shown in Fig. 11. These are the results of work by Professor Woschni at Munich. Many other programmes of a similar nature have been, and are being, developed.

When the limit of thermal stress is reached the only method of further increasing the engine output is by the adoption of an increased air/fuel ratio and hence shifting the onus to the turbocharger designer.

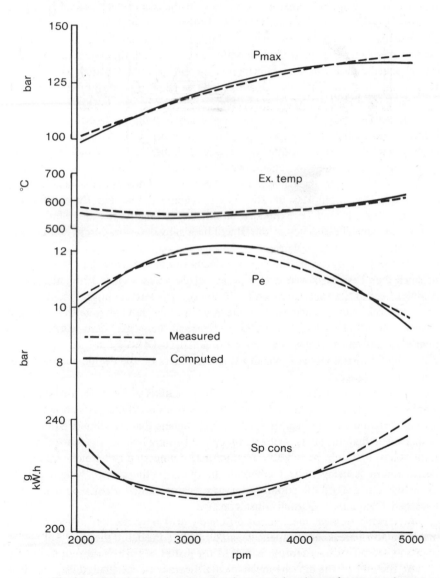

Fig. 11 – Comparison of measured and computed engine performance values.

4. MOTORING LOSSES

A major problem in high speed diesel engine design is to ensure that the mechanical efficiency is as high as possible. Mechanical efficiency is limited by the high frictional losses with which the diesel engine is plagued as a result of the high cylinder pressures of the cycle. By comparison with a gasoline engine of a similar size the diesel engine will have greater bearing areas, a heavier piston, heavier connecting rod, greater piston ring loadings and very probably higher parasitic losses in the form of coolant and lubricating oil pump drives. The sum of these frictional losses does not rise linearly with speed but at an increasing rate and hence tends to limit the maximum speed at which the engine can operate. At all speeds, motoring losses influence the fuel economy which is the chief reason for the diesel engine's existence. It is worth a great deal of effort, therefore, to reduce friction wherever possible.

Figure 12 shows how total engine friction is built up from the frictional losses of a number of components. The losses are typical of swirl chamber engines and the actual loss due to the swirl chamber itself is shown separately. Clearly, the direct injection engine would have the advantage of much reduced losses associated with the combustion process itself and would also benefit from the use of a lower compression ratio. The problem is to maintain the combustion efficiency (which may be expressed as air use) at high speed as otherwise the gain in mechanical efficiency of the DI will be lost by deterioration in combustion efficiency.

Very little progress has in fact been achieved in reducing the motoring loss of small diesel engines, even over a period of 20 years. Figure 12 highlights the Achilles heel of the small diesel and if in the future there are improvments in the economy of gasoline engines, as there may well be now that both users and designers are 'fuel efficient' conscious, the small diesel may find it difficult to justify its existence. It may be that the high speed DI will come to the rescue, but if the DI carries any cost penalty this will reflect on the overall economics of the engine package.

In order to minimize bearing areas, the crankcase must be stiff to limit crankshaft deflection. A main bearing between each cylinder is strongly to be advised. Greater skill in bearing load analysis means that margins of safety may be narrower than in the days of rule of thumb bearing loading figures and by the same token it should be possible to minimize connecting rod weight by refined design methods using finite element techniques. In the piston, weight can be saved by the use of oil cooling at lower engine ratings than would make it necessary from purely thermal considerations.

An engine of high mechanical efficiency will only result from meticulous attention to all areas that influence engine friction. Probably the area that would pay most from further attention is that of the piston and piston ring combination.

At the end of the day the mechanical efficiency of the small diesel engine is likely to be some 40% or more lower than that of its gasoline engine counterpart;

this is for car and truck applications. In larger engines there are, of course, few gasoline engine counterparts, and as engine ratings rise and increasing levels of supercharge are employed, the effect of mechanical efficiency diminishes.

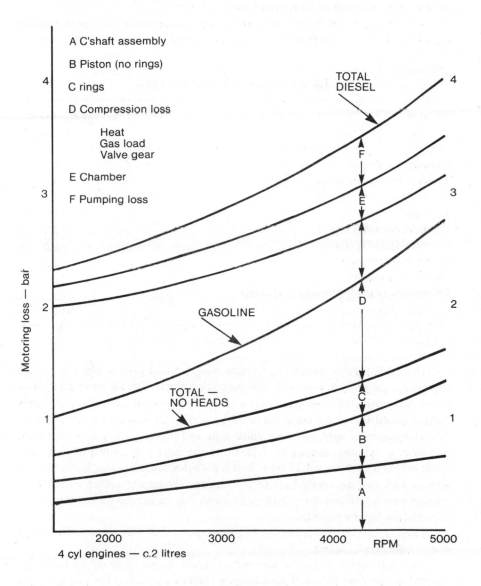

Fig. 12 – Engine friction.

5. OVERALL DESIGN

Weight and overall size are of the greatest importance in diesel engine design. Reduced engine weight not only saves direct cost in the engine itself but also there are down-the-line savings in any vehicle application in reduced cost of the enclosure and the braking system, etc. Equally, reduced size will permit the use of any one engine in a wider range of applications.

Table 1 shows the weights of various engine components as a percentage of the whole. These are for four and six-cylinder in-line engines.

Table 1. Engine component weights

Component	Percentage of total
Cylinder head	8
Crankcase/cylinder block	34
Crankshaft	9
Flywheel	11
Piston assemblies (set)	2
Connecting rods (set)	1.8
Injection equipment	4.2
Starter + generator	10
Other parts (each less than 2% of total)	20
	100

This table may be applied to engines from 1.5 to 4 litres swept volume and engines should show a weight of less than 100 kg/litre. The figure of 80 kg/litre would be considered very good and a large number of commercially available engines would fall in the range 80 to 95 kg/litre. It has already been mentioned that dimensions of approximately equal bore and stroke are to be preferred. On this basis a cylinder spacing of 1.18 times the bore will comfortably permit water between each bore. 1.15 times bore is probably the minimum. For air cooled engines such spacing would have to be considerably greater or the restricted air passage way would impose considerable losses due to the fan power required to drive the cooling air through.

Table 2 lists typical other leading dimensions that will control the overall size of high speed engines.

Any high speed engine of current or recent design is likely to be turbo-charged in the near future. Turbocharging offers a great deal for diesel engines but the development of small turbochargers to suit engines to 3 litres has taken a

Table 2. Leading dimensions of engine

Crank pin	
Diameter	60% bore
Length	33% bore
Crank journal	
Diameter	69% bore
Length	35% bore
Crank web width	23% bore
Connecting rod centres	1.6 X stroke
Gudgeon pin diameter	33% bore
Piston	
Compression height	52% bore
Skirt depth	40% bore
Piston/cylinder head clearance	1% stroke

long time. Perhaps it was not until the turbocharging of small gasoline engines became fashionable that sufficient manufacturing volume was seen to justify the development of small turbochargers.

In fact, provisions necessary to ensure that a turbocharger may readily be fitted are not difficult. Less so than those necessary for the conversion of an IDI engine to a DI engine.

The configuration of the inlet and exhaust ports is a matter for debate. In general, a cross flow head with inlet and exhaust ports on opposite sides is to be preferred, but for an in-line engine the turbocharger installation is simplified if all the ports are on the same side. For vee-form engines it is nearly always preferable to arrange for the inlet ports to be on the inside of the vee and the exhaust ports on the outside. As regards the installation of a turbocharged engine in a vehicle the turbine casing and the exhaust pipes converging on it represent an unwelcome hot-spot. The need to accommodate this and perhaps to provide some controlled air movement across it may conflict with other requirements for noise limitation by enclosure.

There are, of course, problems of matching the output of a turbocharged engine to road load requirements and a waste gate will in many cases be necessary. In spite of this, and even with some boost control of injection, matching is often less than ideal as may be seen from tell-tale smoke puffs on acceleration.

Probably the most important consideration that the designer of a new engine will have in mind is to minimize the noise that will be radiated from the engine's structure. Much work has been done on this subject and is continuing, and progress in noise reduction has certainly been made, though it can hardly be described as dramatic. It is still quite easy to detect with the ear the difference between diesel engine and gasoline engine passenger cars just as it was 20 years ago. Perhaps gasoline engines have got quieter too.

Three references in this group bear directly on design for noise reduction and others deal indirectly with the matter, and so comments here are brief. Approaches that are advocated by workers in the field are to try and keep the noise inside the engine. This may be done by encapsulating the engine in its own enclosure and by providing it with inlet and outlet ventilating ducts or by effectively enveloping the engine in its own wrap-round sump. Both these approaches have their disadvantages. The first seems relatively expensive and the figure of 10% increase of the complete chassis price has been quoted as the cost of this method of installation. The second method involves very radical changes in the basic engine design. The most cost effective approach for noise reduction is undoubtedly by attention to the crankcase design, particularly with regard to main bearing restraint and attention to all other detail aspects such as piston clearances, timing drives and engine covers.

In the U.K., regulations concerning noise have not to date been very rigorous but they are now (1983) beginning to tighten. Although requirements may still be met by conventional and relatively inexpensive methods at present, there is no doubt that in future much more drastic methods, probably involving encapsulation, will become mandatory.

It is easier to apply these noise limiting principles to liquid cooled engines than to air cooled engines. Although there are very successful highly rated air cooled engines in production today, the road to success is so hard that few other manufacturers seem willing to attempt it. It would in any case be impossible to deal with the particular design requirements of both types of engine in this short section and therefore comments can only be confined to generalities.

The vast majority of engines to be introduced will fall into the configurations of in-line four or six and V6 and V8. Illustrations of representative designs of these are shown in Figs. 13, 14 and 15.

6. TIMING DRIVES

The early days of small, high speed diesel engine development were dominated by valve gear problems. At that time there were very few overhead camshaft gasoline engines in large scale production and these were mostly confined to high performance cars. Engines for utilitarian vehicles, both gasoline and diesel, almost invariably employed push rod and rocker operated valve gear and with the narrow valve timings that are obligatory in small diesel engine cylinders it was a difficult compromise to achieve adequate lift and satisfactory contact stresses. The problem stimulated work on cam design and drew attention to the great importance of stiffness for the components making up the valve train.

84 x 82

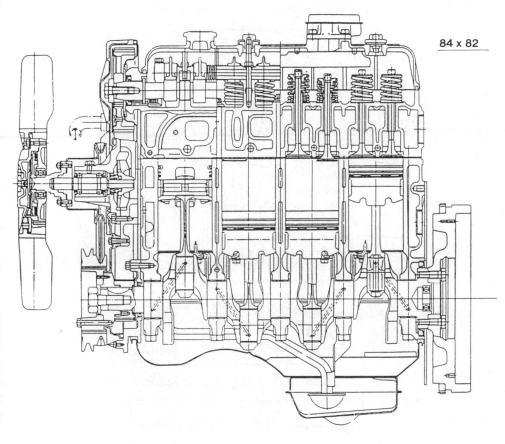

Fig. 13 – Representative design of 4 cyl. in-line engine.

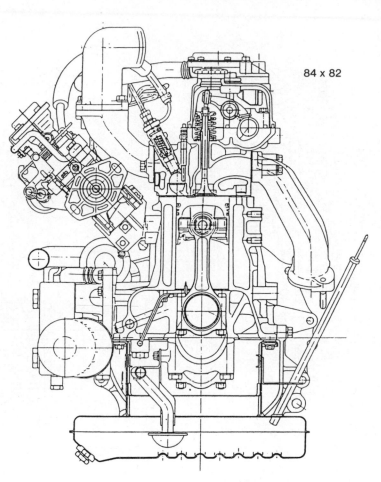

84 x 82

Fig. 14 – Cross section of engine of Fig. 13.

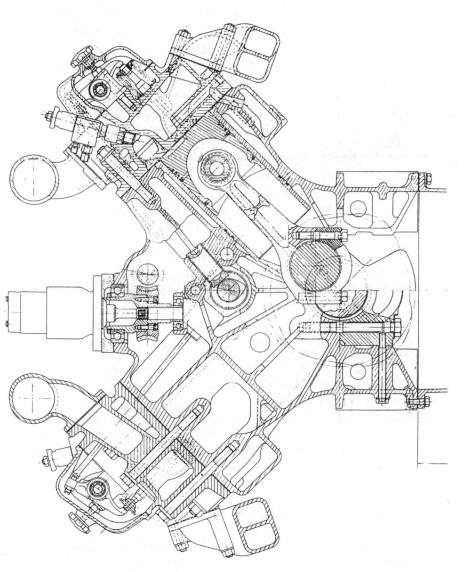

Fig. 15 – Representative design of V8 engine.

Push rod engines are still with us but the arrival of the timing belt and the development of the belt to the extent that it can also satisfactorily drive the injection pump, at least for IDI engines, greatly simplifies the engine design. For a new engine design, that is to say where there are no constraints imposed by existing production machinery, the first choice would be an overhead camshaft driven by a timing belt which would also drive the injection pump. Even the back of the belt can be used for driving some auxiliaries. This solution should be suitable for engines up to 500 or 600 cc per cylinder.

Where for size or other considerations the timing belt is not acceptable as a camshaft drive, duplex roller chain is often the best solution. This will, in practice, mean a push rod engine. The relative positions of the crankshaft, camshaft and injection pump will determine whether a triangulated drive can be employed or whether two distinct drives are necessary.

In either case it is important to keep the chain drives short, as chains wear and this causes the injection timing to retard. Chain drives are extremely tolerant, will put up with the gyrations of the front end of the crankshaft and are reasonably quiet in operation. Their use is quite satisfactory for in-line engines of up to at least six cylinders and 900 cc per cylinder.

For vee-form engines it is much more difficult to accommodate a roller chain drive. The use of a multiplicity of timing belts is also not very welcome. Here, therefore, there is much to be said for a spur gear drive. It is therefore difficult to consider an overhead camshaft for such engines but the camshaft can at least be positioned relatively close to the cylinder head and hence push rods may be kept short.

In push rod engines the camshaft may operate tappets which are either of the bucket type or of the mushroom type. Generally, the latter can be accommodated more easily without increasing the width of the engine. For an overhead çamshaft design it might be thought that once the step had been taken of putting the camshaft in the head it should operate the valves directly. Direct attack is undoubtedly the best solution for very high speeds but diesel engines do not operate at very high speeds and many designers favour the use of a rocker in overhead camshaft engines, chiefly for ease of valve clearance adjustment.

7. CYLINDER HEAD AND GASKET

Problems of making a satisfactory seal with a detachable cylinder head have led many designers to advocate the use of an integral head, particularly for DI engines at high ratings. Experience with examples of those that have actually appeared has not encouraged many others to follow this course. Gaskets, with their attendant sealing problems, are likely to be with us for a long time.

Maximum cylinder pressures are closely related to specific output. Fig. 16 shows how maximum pressures increase as the supercharge pressure ratio is

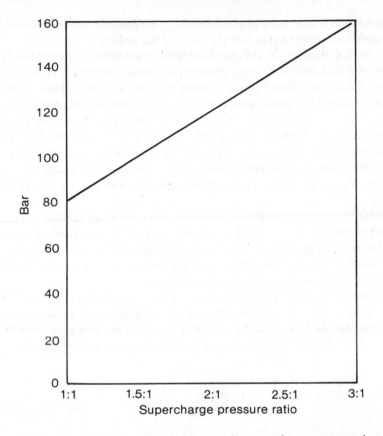

Fig. 16 – Maximum cylinder pressure against supercharge pressure ratio.

increased, and this curve as presented allows for some reduction in compression ratio as supercharge rises in order to minimize the effect on pressures.

In small engines for vehicle use there will be a strong urge to adopt a pattern of four bolts per cylinder, as this is normal gasoline engine practice and can provide a symmetrical and simple layout. For a swirl chamber engine such an arrangement is not ideal, as the swirl chamber itself may extend outside the bolt line. However, alternative patterns are not ideal either and if more bolts are used they may influence the shape of the valve ports adversely. It is well worth trying to retain the four-bolt pattern and to arrange for stiffening ribs within the cylinder head and under the cylinder block top deck in order to minimize local deflection.

Passages between the cylinder block and the cylinder head for coolant, oil and push rods are unwelcome as they weaken the metal faces. It is generally possible to locate the coolant and oil passages away from the gas sealing area and this should be done as far as possible. Also such passages should be sealed by lightly loaded grommets and not by flat areas of the gasket.

The load applied by the cylinder head bolts or studs should be not less than 3.5 times the load due to the firing pressure, and should preferably be 4.5 times. The head should be designed to concentrate the load around the bore or, in the case of swirl chamber engines, around the bore and combustion chamber area.

Figure 17 gives an indication of the sort of valve sizes that may be adopted with various types of engine. The diesel example is a two-valve swirl chamber one. Because of the narrow valve events there are generally limitations on the valve lift that can be used with diesel engines, and hence breathing can be a limitation. Figure 18 shows valves overlapping the bore, which is a possible expedient.

For IDI engines it is normal to employ a single inlet and exhaust valve, and the point has already been made that it is quite possible to obtain adequate valve area without forming a dangerously narrow bridge between the valves. For DI engines the position is quite different, as the injector encroaches on the space between the valves and the valve sizes are restricted.

Four-valve, two inlet and two exhaust, configurations have often been used for DI cylinder heads, and for naturally aspirated engines of above, say, 125 mm bore, there may well be a case for adopting this solution. However, there is of

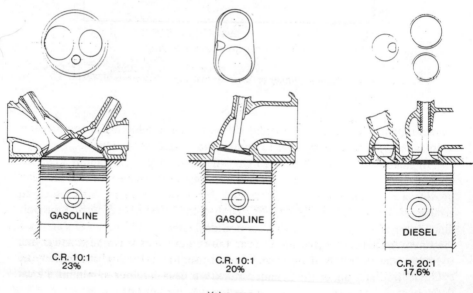

GASOLINE   GASOLINE   DIESEL

C.R. 10:1   C.R. 10:1   C.R. 20:1
23%   20%   17.6%

Valve area
Figures show — inlet valve throat area
Cylinder bore area %

Fig. 17 – Valve sizes adoptable for various types of engine.

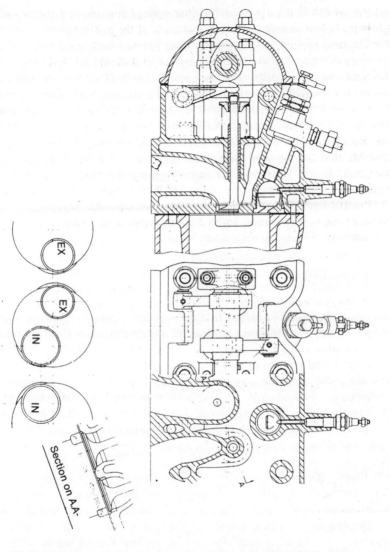

Fig. 18 – High speed diesel engine with valves overlapping cylinder bore.

course a cost penalty in the use of four valves and the cylinder head design can become very congested. If the engine is to be turbocharged, and increasingly all such engines will be, it is less easy to make a case for four valves, as the air velocity at the inlet valve throat is not necessarily the limiting factor.

For IDI engines with valve sizes of perhaps 45% and 37% of the bore for the inlet and exhaust respectively, all that is necessary is to develop free flowing port forms. The technique of flowing the ports on a rig at a fixed pressure drop and varying valve lift is quite valid for developing optimum port forms within the confines of the cylinder head in question. It is worth making a number of wooden models of port forms to determine the optimum in this way.

For DI engines where the inlet port also provides air swirl there are two problems: first, that of designing a suitable port form and secondly, that of holding it consistently in production. Port forms as cast are likely to vary, even in one cylinder head let alone between engines. For this reason, fully machined inlet ports have been investigated but it is extremely difficult to achieve as good a level of swirl with machined port forms as with the best cast forms.

Rig swirl rate tests of inlet ports have long been established by arbitrary swirl meters, but more recently laser doppler techniques have led to greater refinement by permitting tests on motored or running engines. Probably LDV techniques should be considered for the development stage of any new DI engine of high performance, and a suitable window or windows provided in the cylinder head. This is not always an easy task.

8. WHAT'S NEW?

8.1. Fuels–Conventional

It seems to be generally agreed that the quality of gas oil is going to deteriorate because of the demands on the supply of middle distillates. Cetane number may well drop from 50 to 45 and there may be an increase in viscosity and changes in volatility. Diesel engines as they are today have been developed to suit fuels as they are today and thus any changes in fuel specification will require engine development. Such changes, as have been mentioned, will have more effect on DI than on IDI engines. Even so it should not be difficult to keep pace with changes in fuel specifications as foreseen at present, though changes to injection equipment design and swirl requirements will be necessary.

8.2. Fuels – Alternative

A different challenge will be presented by increasing demands to operate engines on alternative fuels, of which the most likely are vegetable oils and alcohols.

Developing countries, fearful of the loss of petroleum products if some emergency should arise, naturally want to provide for the use of some proportion of home grown oil to be sure of being able to harvest and distribute their crops. Many vegetable oils differ little from gasoil and although they are not suitable for use untreated in diesel engines, relatively simple chemical manipulation will render them perfectly acceptable. At the same time some quite simple modification of current engines will enable them to use such vegetable oil to maximum advantage.

A more difficult case is the use of the alcohols. In some countries ethanol (ethyl alcohol) will become available for fuel use from the fermentation of biomass followed by distillation. In other areas methanol (methyl alcohol), which can readily be made from natural gas or any other carbonaceous feedstock, will be available for use as fuel.

Neither alcohol is at all suitable as a diesel fuel but a small proportion of either can in fact be combined with vegetable oils in a process called esterification to yield a very satisfactory diesel fuel.

What are the possibilities for the use of the alcohols unmodified?

Not only do the alcohols have very poor ignition qualities (methanol is about 0 cetane no.) but it also has very low lubricity, which is bad from the point of view of the injection pump and nozzle, and a very high latent heat which means that the temperature of the air charge will be sharply reduced when the alcohol is injected into it. Methanol also attacks many materials, particularly if mixed with water to any degree.

To raise the cetane number of methanol to an acceptable level would require a quite uneconomic proportion of cetane improver such as cylohexylnitrate. The same is true of castor oil, although this would improve the lubricity.

Work has been done on 'fumigation' by admitting methanol to the manifold in atomized form by an injector. Although little success was achieved in the past, the approach seems worth reviving because of its simple nature.

A successful method is to use dual injection with a pilot charge of gasoil entering the combustion chamber first followed by a 'power' charge of methanol from another injector. It is, of course, a very expensive method.

The use of a 'hot spot' to ignite the alcohol with, for instance, a glow plug would take advantage of its tendency to pre-ignition. This is perhaps worth investigation.

All in all there is plenty of scope for work on the alcohols.

8.3. Electronic Injection Systems

Much work is in progress on the development of electronic injection systems, and commercially available systems have been announced. They may take the form of full authority control systems or provide a measure of trimming to the injection characteristics. In theory, electronic control should provide for closer regulation of the shape of the injection diagram which will become of increasing importance as standards of diesel engine combustion and noise rise but it is not yet certain that this can be achieved at reasonable cost.

8.4. Higher Outputs

More will be demanded from diesel engines, and increased output will only come from supercharging. Perhaps it will be essential to live with higher cylinder pressures. This will certainly put a premium on mechanical design, particularly as regards the gasket joint and the bearing loadings.

Economy, too, will be sought. Marine diesels operating at low speed and with high supercharging return a fuel consumption of 125 gr/CV/hr, which is to say an overall thermal efficiency of over 50%. This is a target to aim for.

8.5. Cooling Systems or the Lack of Them

It is unlikely that we have heard the last of the 'adiabatic' or 'low heat loss' engines. Engines are running and the principal advantage for them is perhaps the elimination of the cooling system. This is a considerable prize. Unfortunately the necessary development work on any particular engine or family of engines is likely to be extremely expensive and probably few are brave enough to undertake it.

Work that has been done on the use of ceramic components of low thermal conductivity to minimize conduction of heat from the combustion chamber and cylinder has shown the need to treat all relevant surfaces. If some surfaces are formed of low conductivity ceramics and some of metal then the metal will provide a path for greater heat flow and the net result as regards heat flow to coolant will be much the same as if all the surfaces were metal. Hence if ceramics are to be used they must be applied to the cylinder head face, the valve heads, the piston crown and the liner.

There is no doubt, however, that more knowledge concerning heat flow in relation to thermal stresses and cooling systems would pay dividends, and there is plenty of scope for work here.

2

Design of high speed diesel engines

J. N. Manning, Ricardo Consulting Engineers plc

1. INTRODUCTION

The diesel engine has long been associated with heavy commercial vehicles where the gain in fuel economy and longer overhaul periods of the diesel compared to the gasoline engine were highly attractive.

During the 1950s there was a significant increase in the number of small, high speed four-stroke diesel engines manufactured, particularly in Europe, for use in agricultural tractors, light trucks, delivery vans and taxis. These again were applications where the economy of the diesel engine was attractive and its disadvantage of lower specific power output and increased weight and noise, compared to the gasoline engine it replaced, were less important than the gains in economy, reliability, and longer overhaul periods, particularly under start–stop operating conditions. From the 1960s onwards the application of the high speed four-stroke diesel engine has expanded into the passenger car market, accelerated by the world fuel crisis and the resulting desire to conserve energy reserves.

The incursion of the high speed diesel engine into the field of passenger cars has demanded from it ever higher standards in order that it may compare favourably with its gasoline counterpart not only in terms of performance, weight and noise but also in manufacturing cost and marketing requirements. These higher standards inevitably pose a greater challenge to the diesel engine designer who, in addition to the traditional considerations of economics and applications, now also has to be concerned with legislative requirements which are exerting a growing influence on engine design.

All these factors have contributed to the evolution of a new generation of high speed automotive diesels and it is therefore proposed to concentrate here on the design of these engines, although much of the design theory applies equally well to the heavier duty diesel engines for the more traditional automotive and industrial applications.

2. OBJECTIVES

Before commencing the design of an engine it is necessary to define the objectives and these can be broadly grouped under the following headings:

1. Application.
2. Performance.
3. Manufacture.
4. Product Quality.

The effects of each of these aspects on the design concept are now considered.

2.1. Application

The high speed diesel engine for automotive applications is invariably subject to more severe installational constraints than in other potential applications such as the industrial and marine engine markets. The two main installational aspects to be considered are package size and engine weight, both of which are of increasing importance in the passenger car which is being progressively down-sized in the quest for improved operating efficiency.

Assuming that the installation requirements have been largely defined by vehicle engineering and styling, the choice of engine configuration may be limited. A length limitation could restrict the number of cylinders of an in-line engine and favour a vee formation or even horizontally opposed cylinders. However, the resulting length reduction may not be significant compared with the increase in width, which can be a major disadvantage for a horizontally opposed engine in a front installation where the space between front wheel arches is limited. This can have further implications in accessibility for servicing. In comparison the simplicity and small 'box volume' of an in-line engine, particularly one with four cylinders, make it very adaptable for many installations, both longitudinal and transverse. In addition it can be inclined at any convenient angle to suit a particular requirement, such as a low bonnet line, which is of increasing importance with the attention currently being paid to vehicle aerodynamics.

When choosing a cylinder layout consideration must also be given to engine vibration and firing interval, since the number of cylinders and the angle between the cylinder banks on a vee engine will influence the external vibratory forces produced. These disturbances are of two specific kinds: (1) unbalance due to the reciprocating masses, i.e. piston, piston pin, connecting rod small end, and (2) reaction to the torque produced by both the gas pressure acting on the piston and the inertia of the reciprocating masses.

Considering the first type of disturbance. It is found that no in-line engine of less than six cylinders, and no vee engine of less than eight cylinders, is inherently free of primary and secondary unbalance, although in some cases the vibrations can be satisfactorily absorbed by careful design of the flexible engine mountings. As the number of cylinders decrease a balancer shaft system becomes desirable or essential in order to achieve acceptable smoothness, but such a system can add significantly to the cost and weight of the engine, thus tending to negate these and other friction and cylinder size advantages of an engine with fewer cylinders. The most popular configuration for passenger car engines, four cylinders in-line, has complete primary balance but the secondary inertia forces produce a vertical vibration with a frequency of twice engine speed. These secondary vibrations are worse in the diesel engine than in a gasoline engine of corresponding size owing to the increased weight of the reciprocating parts, and the degree to which these are noticeable within the car will depend on the effectiveness of the engine mounts. These out-of-balance forces can be counteracted by fitting twin balancer shafts contra-rotating at twice engine speed, a solution adopted in many diesel engines fitted to agricultural tractors where the engine is not isolated from the vehicle chassis.

The second type of disturbance, torque reaction or torque recoil, which causes the engine to vibrate in a rolling mode, is a more difficult problem to solve. Torque reaction due to inertia can be controlled by the use of balancer shafts, in the case of the four-cylinder in-line engine by using the same balancer shafts already mentioned but mounting them at different heights. However, torque reaction due to the gas pressure on the piston is the more serious problem on the diesel engine, particularly at idle speeds, because unlike the gasoline engine the diesel is unthrottled and the torque fluctuation is not very load sensitive. The torque reaction due to both inertia and cylinder gas pressure may be counteracted by a method which replaces the engine flywheel with two separate flywheels, one directly mounted on the crankshaft and the other, which has an effective inertia equal to the total inertia attached to and driven by the crankshaft, gear driven from the crankshaft in the reverse direction. The effect of this arrangement is that the reaction to the accelerating torque on the second flywheel, which is taken through the engine structure, cancels the torque causing the recoil. This method of balancing torque reaction can be applied to an engine with any number of cylinders, although the greater the number of cylinders the less severe is the torque reaction problem. As the trend to down-sizing engines for fuel economy continues there is growing interest in the three-cylinder in-line engine as a means of maintaining a reasonable cylinder size for good combustion characteristics. This results in more severe balancing and torque recoil problems which must be overcome if an acceptable vehicle comfort level is to be maintained.

For a given engine configuration and comparable swept volume, the diesel engine will invariably be heavier than a gasoline engine. This increased weight is due largely to the fuel injection equipment and starter motor which are heavier than the equivalent carburettor, ignition system and starter of the gasoline engine. The heavy duty starter motor to cope with the increased compression pressure of the diesel will, in addition, usually require a heavy duty battery. Regarding the engine structure, the additional weight of the diesel can also be attributed to an increase in height of the cylinder block and/or cylinder head depending on the combustion system chosen, and also to some extent the generally more robust components to withstand the higher cylinder pressures. Much attention must be given to the detail design of all engine components to minimize weight by the use of light materials and careful stress analysis. The increasing use of finite element techniques offers the designer greater potential for optimising the design of the more complex engine components such as the crankshaft, crankcase and cylinder block, but of course such measures can equally well apply to the gasoline engine.

If a gasoline engine is to be replaced by a diesel engine of equal power, some increase in weight and package size can be expected, depending on the method chosen to obtain the power. This is not necessarily a problem except when the diesel engine is required to fit in place of a gasoline engine in a vehicle whose engine compartment was tailored to suit the latter.

One further aspect of the engine application which must be considered is the choice of air cooling or water cooling. Air rather than water cooling tends to be chosen where the environment makes it more suitable, such as engines for use in cold climates, or for military vehicles and industrial uses where the advantages of a complete self-contained power system are important. Air cooled engines eliminate the potential liquid coolant problems of freezing, boiling, leaking and corrosion. However, air cooled engines have significant disadvantages compared to water cooled engines, such as supplying and distributing the cooling air which unless great care is taken in the design of the fan and associated ducting, can absorb a large amount of engine power, generate excessive noise, and present a packaging problem. An air cooled engine runs at higher temperatures than a liquid cooled engine and is more susceptible to localized thermal problems owing to the difficulty of controlling the cooling air flow. The lower thermal inertia of air compared to water means that an air cooled engine warm-up is quicker but so also is its cooling down. This and the less stable operating temperature pose problems when considering the needs for heating the passenger compartment of a vehicle, together with the difficulty of handling and distributing air uncontaminated by engine fumes.

In addition to the above disadvantages an air cooled diesel engine for a passenger car application would require higher technology, would cost more per unit power output and would be less adaptable for packaging than an equivalent water cooled engine. Hence the water cooled engine dominates the diesel passenger car market.

2.2 Performance
The power output of an engine depends on its volumetric efficiency, thermal efficiency, and mechanical efficiency. The high speed automotive diesel engine has a lower power/litre and a higher weight/litre than the corresponding gasoline engine, giving a resultant power/weight ratio of perhaps as low as 60% of its

gasoline counterpart. It is therefore important to consider how each of the efficiency factors may be optimized in the diesel engine.

2.2.1 Volumetric efficiency

The volumetric efficiency is determined by the engine's breathing characteristics — its ability to draw air into the cylinders and expel exhaust gases — and the air utilisation once it is within the cylinder. The lower power/litre of the automotive diesel engine is due partly to the lower air utilisation resulting from in-cylinder mixing rather than drawing in a premixed charge, a certain portion of the air present for mixing being inaccessible in the working clearances within the cylinder. The high rate of fuel/air mixing required in the diesel engine requires high air swirl and compression but the associated pumping losses tend to limit the maximum useful speed of the diesel engine which is therefore lower than that of the corresponding gasoline engine. A useful guide to determine this realistic rated speed is to consider the mean piston speed which for a passenger car diesel has a reasonable rated limit of around 14 m/s, although this is also dependent on mechanical and friction considerations. In addition the relationship between volumetric efficiency and mean inlet valve gas velocity, calculated at the inner diameter of the valve seat, indicates that no further increase in power can be expected by increasing the engine rotational speed once the inlet valve gas velocity has reached about 75 m/s (Fig. 1). The valve area available depends on the valve layout adopted, but with the conventional layout of two vertical valves operating within the cylinder bore then the effective inlet valve inner seat diameter can be around 42% of the bore and the exhaust around 36% of the bore, giving an exhaust gas velocity around 100 m/s at maximum rated speed. The selection of engine rotational speed is a compromise between the demands of vehicle drivability, economy and noise, and a practical maximum rated speed appears to be around 5000 rev/min.

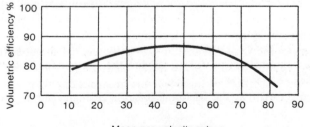

Mean gas velocity m/sec

Fig. 1 — Relationship between observed volumetric efficiency and mean inlet gas velocity.

The stroke/bore ratio for a diesel engine should preferably be greater than unity, that is undersquare, but the requirement for limiting the inlet valve gas velocity in a high speed engine tends towards square cylinder dimensions with the consequent increase in surface to volume ratio and dead volume in the cylinder.

Consideration of these parameters now enables the designer to make a first estimate of engine dimensions and package size.

An increase in engine speed will not significantly increase the specific power of the diesel engine and therefore the alternatives of increasing the air utilisation or the air charge mass in the cylinder must be considered. Scope for improving air utilization is limited by in-cylinder mixing and hence it is desirable to increase the mass of air in the cylinder either by improving the engine breathing or by some type of pressure charging. Improved breathing obtained by optimizing valve timings, lifts and sizes together with careful design of ports and manifolds can offer small benefits in specific power. Pressure charging using induction ram effects could give an increase in power of up to 10% but it is difficult to spread the advantage over the engine speed range and there is the additional installation problem of incorporating long induction tracts. The alternative methods of pressure charging all offer a significant increase in power but at considerable increases in cost and complexity. The three possible systems are turbocharging, pressure wave charging, and mechanically driven compressor.

2.2.2 Thermal Efficiency

Thermal efficiency is somewhat difficult to quantify but the requirements for optimizing it, and hence improving specific fuel consumption, are rapid combustion and minimum heat loss. The former is achieved in practice by ensuring good air/fuel mixing, and the latter by minimizing the surface/volume ratio of the combustion chamber which is assisted by having an undersquare stroke/bore ratio. There is further potential for minimizing heat loss by insulating the combustion chamber and exhaust port, and ceramic materials are currently being assessed for such applications.

2.2.3 Mechanical Efficiency

The mechanical efficiency is a measure of the friction losses within the engine and also of the external losses due to driving the cooling fan, alternator and other accessories.

It has been shown that the frictional losses in an engine can be divided into three groups:

(a) those approximately invariable with speed, which are related to the compression ratio,

(b) those varying directly with speed, which are related to the rotating and reciprocating components,

(c) those varying with the square of the speed, which are related to the pumping losses.

High engine rotational speeds should therefore be avoided in the interest of high mechanical efficiency, but because of its higher compression ratio the diesel engine will always be at a disadvantage compared to the gasoline engine.

2.3 Manufacture

The cost of manufacturing a diesel engine will inevitably be higher than for a gasoline engine. For equal power outputs the diesel engine has a larger swept volume, increased weight, and requires expensive fuel injection eqipment. For equal swept volumes the diesel engine will require some form of pressure charging to produce the same power as the gasoline engine and the additional cost of this will be at least comparable to or possibly greater than increasing the swept volume. In addition the manufacturing tolerances allowable on major diesel engine components such as the cylinder block, crankshaft, connecting rod and piston are tighter than for equivalent gasoline engine components because of the need to control the higher compression ratio and smaller piston to cylinder head clearance used (Fig. 2).

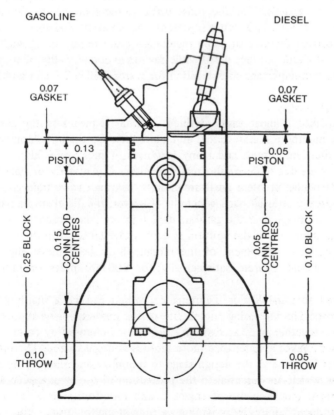

Fig. 2 – Tolerances controlling piston to cylinder head clearance (mm) for engines of 65 to 110 mm stroke.

However, in the highly competitive passenger car market it is absolutely essential to achieve the lowest manufacturing cost which is consistent with meeting the other targets for the engine. Unless the engine is to be a completely new design owing nothing to existing production units it is essential to use existing components and tooling. Frequently the passenger car diesel is based on a current gasoline engine and must therefore be produced on existing manufacturing plant with the minimum adaptation. This can place constraints on cylinder bore, stroke and centre distance, cyliner block height, cylinder head bolting pattern, and auxiliary drives. Components which are common to both gasoline and diesel variants, such as possibly the cylinder block, crankshaft and connecting rod, can be graded after machining such that those falling in the centre of the allowable tolerance band are used for the diesel build and the remainder for the gasoline build.

The possibility of an engine eventually being produced in a family of, for example, three, four, five, and six-cylinder versions to cover a wide range of marketing requirements must be considered at the start of the design process. It is important to design for the most difficult members of the family, where technical or production features present the greatest challenges. This could result in certain features of some members of the family being slightly over-designed but could produce worthwhile savings in commonality of components, which is also an important consideration for spares availability and servicing.

2.4 Product Quality

Product quality is most easily defined by what it means to the customer — reliability, minimum maintenance, and durability. Of these, reliability is probably the most important, and many instances of premature failure of engine components are due to design shortcomings rather than material or manufacturing defects. Particular problem areas are gaskets, bearings, valve train components, and dynamically stressed parts generally. However, the diesel engine has a major reliability advantage over the gasoline engine in the elimination of high voltage electrical equipment for the ignition system. Provided fuel filtration is adequate the fuel injection equipment of the modern high speed diesel engine, if undistrubed, should operate satisfactorily for 100 000 miles or more before overhaul.

Related to reliability is maintenance, which can be a major portion of running costs. The increasing complexity of the modern engine and its ancillary equipment, together with accessibility limitations imposed by compact vehicle installations, makes maintenance more tedious and specialized. Therefore every effort must be made at the design stage to minimize and simplify maintenance procedures which are essential to the protection of the engine, such as oil and filter changes. Increased use of engine health monitoring systems will help to avoid component failures by providing warning of malfunction.

Engine durability is a function of the operating conditions, the quality of routine maintenance, and the integrity of the component design and manufacture. In the case of a new engine design, durability testing of prototypes will reveal many of the weaknesses present. Five hundred hours of mixed-cycle test bed endurance running together with thermal shock testing for cylinder head gasket performance should be considered as a minimum target. It is, however, important to consider engine durability in relation to vehicle durability, particularly in the passenger car market.

Having now defined the design objectives for the new engine it is possible to compile a functional specification from which the designer can work. Before commencing with the mechanical design considerations, however, one further decision must be made — the choice of combustion system.

3. CHOICE OF COMBUSTION SYSTEM

3.1 Indirect Injection Diesel (Fig. 3)
The indirect injection combustion system is used universally in current passenger car diesel engines because of the economic and technical ease with which it can be made compatible with existing gasoline engines. It provides adequate specific power, a good operating speed range, good economy, low exhaust emissions and

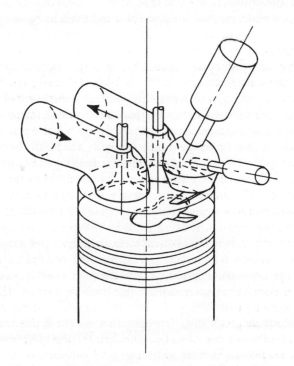

Fig. 3 — Ricardo Comet indirect injection diesel engine.

acceptable noise levels. The most widely used type is the Ricardo Comet swirl chamber.

This system employs a chamber in which fuel is sprayed into a swirling air motion which assists the air/fuel mixing process. Combustion takes place initially in the swirl chamber and then in the main cylinder chamber. The use of a swirl chamber allows operation at relatively low peak cylinder pressures (up to 85 kg/cm^2 for a typical naturally aspirated engine) with low rates of pressure rise, which minimizes cylinder head gasket and bearing loading problems and gives acceptable noise. Low pressure fuel injection equipment (about 350 kg/cm^2) with a variable orifice nozzle can be used, ensuring good metering over a wide speed and load range, with low cost and easy maintenance. The inlet ports can be simple because they do not have to provide swirl.

However, the high thermal losses of the indirect injection system demand a high compression ratio which makes control of combustion chamber volume tolerances difficult in a mass-production environment, and a cold starting aid is required in almost all applications, usually an electric heater plug. Finally, the fuel economy of the indirect injection engine suffers because of the loss of volumetric efficiency resulting from the method of producing air motion.

3.2 'Toroidal' Direct Injection Diesel (Fig. 4)

This system is currently normally used in truck and other heavy duty applications where it offers a 10–20% improvement in fuel economy compared with the indirect injection system. However, the maximum gain in economy is at full load operating condition which represents only a small proportion of a typical passenger car operating cycle.

In the direct injection diesel engine the necessary air motion to achieve good air/fuel mixing is achieved by using a high swirl helical inlet port. The fuel injector position and angle are critical to ensure good full and part load performance and emissions but in practice the constraints placed on the cylinder head layout of small engines often dictate an injector position offset from the centre of the combustion bowl in the piston and excessive inclination from the vertical. In addition space is often limited for the swirl generating inlet port. The relatively low air motion generated in the combustion chamber imposes higher demands on the fuel injection equipment which tends to be expensive because of the very high pressures (greater than 600 kg/cm^2) and small hole nozzles required. The fuel is sprayed through a multi-hole nozzle into the swirling air charge in the piston bowl, the combustion process producing high rates of pressure rise and high maximum cylinder pressures so that noise levels tend to be higher than with the indirect injection system.

The main problem with this direct injection system is the matching of the fuel injection equipment over a wide engine speed range, and possible ways of achieving this are with a variable swirl inlet port, advanced electronically controlled fuel injection equipment, or the use of the unit injector. The last-named

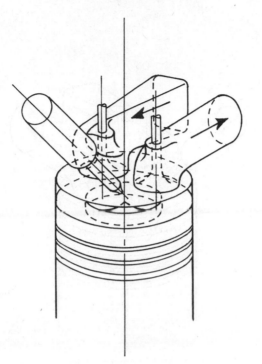
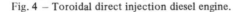

Fig. 4 – Toroidal direct injection diesel engine.

is a self contained pump/injector unit, camshaft activated and able to produce much higher injection pressures than a conventional pump/pipe/injector system. It is, however, at present a very bulky and costly component and is therefore not yet very suitable for small passenger car diesel engines.

3.3 'Wall-wetting' Direct Injection Diesel (Fig. 5)

This system uses a deeper, more spherical, combustion bowl than the conventional toroidal bowl in the piston. The fuel is sprayed into the swirling air charge in the piston bowl but is directed at the wall of the bowl. Combustion occurs partly by the normal fuel droplet burning process and partly by evaporation and burning from the wall of the piston bowl. As with the toroidal direct injection system the layout of the swirling inlet port is important, but because the injector can be angled over to one side of the cylinder head and away from the crowded central area, there is more freedom for port design and positioning, even with the inclusion of a cold-start heater plug. The single hole injector nozzle, ideally with variable orifice, allows low fuel pressures (about 350 kg/cm^2) to be used with resulting savings in fuel pump cost.

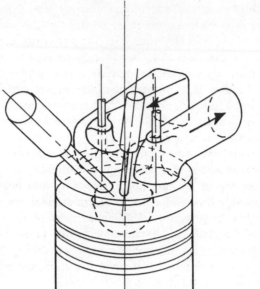

Fig. 5 – Wall-wetting direct injection diesel engine.

4. MECHANICAL DESIGN

The mechanical design of the engine is most conveniently discussed by considering separately the major component sections, although in practice there is a close interrelation between them in determining the overall design of the engine package.

4.1 Cylinder Block

The functions and requirements of the cylinder block can be summarized as:

(a) to provide sufficient overall strength and rigidity to act as the engine frame,
(b) to provide the support and rigidity required in the main bearing housings and top deck,
(c) to provide suitable cylinder bores or location for separate cylinder liners,
(d) be simple to cast,
(e) have low manufacturing cost,
(f) have minimum size and weight,
(g) distribute coolant around the cylinders.

The overall length of a multi-cylinder engine is largely determined by the distance between adjacent cylinder bores, which is now usually determined by the design and construction of the cylinders rather than the crankshaft and bearing requirements which in the past tended to be the limiting feature. Improvements in modern bearing materials have enabled bearing widths to be reduced such that it is now only in the case of horizontally-opposed or vee engine configurations with one crankthrow per cylinder that the minimum bore spacing is usually determined by the crankshaft design. As far as actual cylinder construction is concerned the two forms in general use are the removable wet liner and the integral cylinder barrel with or without a dry liner. The cylinder spacing required with wet liners will be greater than that required with integral barrels if the liners are located at the top of cylinder bores. However, with bottom located wet liners there is probably little difference in centre distance compared to integral barrels provided that the latter are completely surrounded by coolant for at least the upper half of the piston stroke. Some reduction in overall block length is possible if the coolant space between the barrels is deleted but the resulting siamesed cylinders are more prone to thermal distortion, and piston scuffing and oil control problems can occur (Fig. 6).

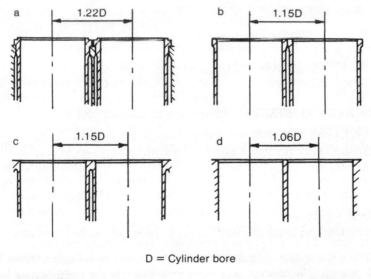

D = Cylinder bore

a Wet liner top location   c Integral bore
b Wet liner bottom location   d Integral siamesed bore

Fig. 6 – Cylinder construction showing normal minimum spacing.

Which type of cylinder construction is adopted will depend on the requirements of the functional specification already defined for the engine. The use of separate liners, either wet or dry, enables a better quality cylinder bore material to be specified as the liner is not subject to the same foundry and hence metal-lurgical constraints as apply to the cylinder block where ease of casting is of prime importance. The liner material can be chosen to give improved wear resistance to the cylinder bore. Dry liners offer advantages in common with an integral bore block of structural stiffness, particularly in the top deck for good gasket sealing, and compact cylinder centre dimensions. They also offer cheap and simple service replacements. Typically, dry liners are of centrifugally cast medium phosphorous grey iron with a wall thickness 2 to 2.5% of the cylinder bore. A light press fit in the parent bore ensures good heat transfer. Dry liners may be flanged at the top or plain, the former providing a positive gasket seal on the flange to prevent gas leakage behind the liner but are more expensive to produce and are prone to flange cracking unless the detail design is correct. Thin steel liners are available as an alternative to cast-iron but the bore needs to be chrome plated to provide a compatible surface. Being chrome plated they are more expensive than cast-iron liners but have very good wear resistance. Wet liners have the following additional advantages:

(a) uniform bore to water jacket thickness and thus better thermal distribution,
(b) thicker castings which are easier to produce and handle,
(c) enable a simple block casting to be used.

However, it is more difficult to achieve a rigid top deck with a wet liner construction and hence good cylinder head gasket sealing. In addition the stresses in the liner seating flange are high and careful design is required to prevent cracking. Attention must also be paid to oil and water sealing, and to prevention of cavitation erosion on the water side caused by liner vibration. A liner wall thickness of at least 7 to 7.5% of the cylinder bore is required, and possibly a protective coating on the outside of the liner. Wet liners with bottom location provide manufacturing advantages because the cylinder block has an open-top water jacket core which is particularly easy to cast. This construction is widely used in automobile gasoline engines with die-cast aluminium cylinder blocks, but has some disadvantages for diesel engines where the top face of the liners must be controlled within close limits to ensure good cylinder head gasket sealing and, with an indirect injection diesel, the uneven loading of the top of the liner due to the combustion chamber insert may lead to bore distortion.

The integral bore cylinder block where the piston runs directly in the main block casting is used extensively in automobile gasoline engines and also light duty high speed diesel engines. It offers advantages of rigidity and compact cylinder spacing and, although the block casting is complex, the overall production costs are low as a result of minimum machining and assembly costs.

The design of the cylinder block must take account of the load path from the cylinder head to the main bearings, with the forces being transmitted as directly as possible through walls and ribs. The cylinder head bolt bosses should be free from integral cylinder barrels to ensure uniform cooling and minimum cylinder bore distortion. The lower part of the block must provide adequate support for the crankshaft via stiff main bearing housings and rigid, well located bearing caps. Lateral location of bearing caps is particularly important on vee engines where there is a transverse applied load and it is necessary to extend the side walls of the crankcase skirt well below the crankshaft centreline. This practice is also beneficial on an in-line engine as it provides a rigid attachment for the flywheel housing and increase beam stiffness generally. An alternative construction to give stiff main bearing housings is the use of a bedplate, or ladder frame, construction. This would probably carry a weight penalty but has been shown to be beneficial in reducing noise radiating from the lower part of the cylinder block.

The detail design of the cylinder block, bearing housings and crankcase is a very complex task if the requirements of low manufacturing cost, light weight, high strength and low noise radiation are all to be successfully met. The only potentially accurate method of anlysing the complex distortions and stresses present is the use of finite element techniques.

The choice of material for the cylinder block is currently between cast-iron and aluminium. Cast-iron has long been the traditional choice as it is an excellent cylinder bore material and offers rigidity, thermal stability and adequate strength. It offers low cost and ease of casting and also provides good noise attenuation. However, compared to aluminium it has a major weight disadvantage, which is increasingly important in the modern passenger car engine. This potential weight advantage of aluminium cannot usually be fully achieved in practice because its lower strength dictates thicker sections, and also there is the penalty of cast-iron cylinder liners, but the potential economic advantages of high volume production die-castings make the use of aluminium for passenger car diesel engine cylinder blocks very attractive.

The use of magnesium alloys offers even greater potential weight savings than aluminium but with the same, if more acute, problems of low strength, high cost, and the need for coolant inhibitors to prevent corrosion in the presence of ferrous components in the cooling system. High volume production is needed for economy.

4.2 Crankshaft and Bearings

The principal objectives in the design of the crankshaft are to achieve satisfactory bearing pressures and material stresses and to ensure sufficient rigidity in bending and torsion.

The most effective way of achieving the necessary bending stiffness, which is important for minimizing noise and vibration, is to reduce the span between the main bearings. This requirement, together with the much higher peak cylinder pressures of the diesel compared to the gasoline engine, has lead to the universal adoption of a main bearing between each cylinder. This has been made possible on the in-line engine without imposing any length increase by the use of high load-carrying capacity shell bearings. Typical bearing materials are overlay plated copper–lead, which has good load-carrying capacity and good fatigue resistance, and 20% reticular tin–aluminium with comparable load-carrying capacity but better resistance to corrosion and cavitation erosion.

A guide to the safe limiting bearing pressures for these materials is 350 kg/cm² for the main bearing, and 550 kg/cm² for the crankpin taking into account the maximum cylinder pressure and the effect of piston and connecting rod inertia forces. However, the calculated bearing pressures are not the most reliable guide to bearing performance, particularly for main bearings, and it is essential to perform a bearing oil film thickness analysis. This will also provide information on the optimum positions for oil drilling break-out points on the crankpin and main journal, and determine whether fully grooved main bearing shells are acceptable, which simplifies the oil feed to the crankpin, or if the required oil film thickness is only attainable with an ungrooved lower half bearing shell, which will necessitate a cross-drilled main journal to provide a continuous pressurized oil feed to the crankpin.

The requirement for balance weights on a crankshaft depends on the engine configuration but the purpose is to eliminate or reduce any external unbalanced force or couple, and to reduce the internal rotating loads applied to the main bearings. Once again the bearing oil film thickness analysis will assist in optimizing the balance weight design with regard to bearing loadings.

Crankshaft torsional rigidity is another important design requirement and vibration amplitudes must be limited to about ±0.25° at the free end of the crankshaft to avoid excessive stresses and possible timing drive problems. The longer the crankshaft the worse is the torsional vibration problem and a damper is usually essential on, for example, a six-cylinder in-line engine. This now typically consists of a hub and a concentric outer inertia ring with a thin cylinder of rubber sandwiched between them, the outer ring also acting as the crankshaft pulley for the vee belt accessory drives. This type of damper has economic advantages over the hydraulic type for high volume production light duty engines.

The stresses in the crankshaft due to bending and torsion must be calculated in the crankwebs and crankpins, taking account of stress concentration factors, at a range of engine operating speeds. These must include idle speed, where in a diesel engine the cylinder gas load is high but the inertia relief effect is small maximum torque speed, full load rated speed, and overspeed conditions.

The choice of material for a crankshaft is between forged steel and spheroidal graphite (S.G.) cast-iron. The latter offers similar bearing performance to steel of comparable hardness but the wear resistance can easily be increased by surface hardening treatment. In addition the fatigue strength of an S.G. cast-iron crank shaft at the crankpin fillet/crankweb/main bearing fillet junction, which is often the most highly stressed region of the crankshaft, can be greatly improved by fillet-rolling. These methods of improving the performance of S.G. cast-iron, together with the lower manufacturing costs resulting from simpler tooling and less machining, have caused S.G. cast-iron to largely replace forged steel for high volume light duty diesel and gasoline engine crankshafts. The casting process also offers the possibility of coring out main bearing journals and crankpins, together with oil feed passages, and thus reducing crankshaft weight and machining time.

Typical crankshaft proportions for an in-line engine, representative of either forged steel or S.G. cast-iron shafts, are shown in Fig. 7.

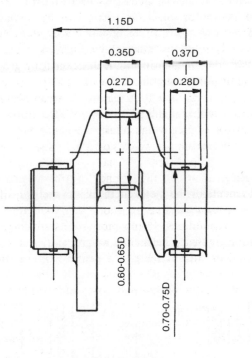

D = Cylinder bore

Fig. 7 – Typical crankshaft proportions.

4.3 Connecting Rod

The design of the connecting rod must logically be considered as part of the crankshaft assembly, particularly the big-end and hence crankpin diameter which must take account of the need to withdraw the connecting rod through the cylinder bore for dismantling. The bearing cap joint should preferably be at 90° to the rod axis as this offers more rigidity, better bearing support, lower cost and easier fastening than the oblique split design. This does, however, limit the crankpin diameter to about 65% of the cylinder bore but this is usually adequate.

The applied load in the cap bolts must first compress the bearing shells into the connecting rod bore and then exceed the maximum inertia force by a suitable safety margin, typically 2:1. The use of bolts and nuts for fastening the cap is preferable to setscrews because their longer length reduces the stress range under dynamic loads. However, an oblique split cap design will require setscrews.

Stressing of the connecting rod shank is straightforward as it is usually sufficient to consider only the direct compressive and tensile stresses due to the gas and inertia loads. However, stressing of the small-end and particularly the big-end requires more sophisticated techniques to allow for the changing section areas and stress concentration factors. The need to minimize connecting rod weight requires more accurate stressing and the application of finite element techniques can enable optimum designs to be achieved.

S.G. cast-iron has for some time been challenging the traditional forged steel as a connecting rod material. The cast rod offers closer weight control because it eliminates the problem of forging die wear during a production run and hence requires less machining. However, both these processes are now being challenged by powder forging which offers excellent mechanical properties and considerably reduced machining.

4.4 Piston and Rings
The main requirements of the diesel engine piston are high strength, low weight, good heat conductivity, good wear resistance and good guidance in the cylinder under all operating conditions. Piston proportions are important for function and also in relation to engine height and weight, and typical passenger car diesel piston proportions are shown in Fig. 8, together with a typical gasoline engine

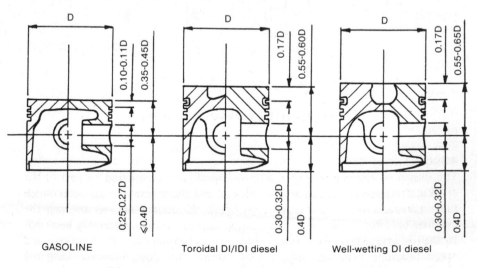

Fig. 8 – Comparison of gasoline and diesel piston designs.

piston for comparison. The compression height, from the piston pin centreline to the crown, affects overall engine height and the minimum acceptable dimension is determined by the piston pin diameter and the specification of the ring pack and its ability to operate reliably. The depth of the top land must be sufficient to prevent excessive temperatures at the ring grooves and the resultant ring sticking problems, and the depth of the second land is determined by the stresses due to the cylinder pressure acting on it. Piston pin diameter is determined by the gas and inertia loads, and the pin wall thickness by the maximum allowable ovalization. Typical maximum piston pin bearing pressures are about $850\,kg/cm^2$ for the small end bush and $550\,kg/cm^2$ for the piston bosses for the preferred fully floating design. The piston skirt serves the function of guiding the piston in the cylinder and its length must be sufficient to ensure steady running which will give good ring sealing and oil control, and to reduce piston slap caused by the lateral loads on the piston at the top of its stroke. The latter problem, which causes piston noise, can also be reduced by offsetting the pin axis slightly from the piston centreline.

In high speed engines the piston weight must be kept to a minimum in order to reduce the reciprocating load, the balance weight mass, the vibration levels, the friction, and the overall engine weight. It is therefore important to keep piston height to an acceptable minimum, and in this context it should be noted that the spherical combustion bowl in the piston of a 'wall-wetting' direct injection diesel engine may dictate a larger compression height than a conventional direct injection or indirect injection diesel piston.

Two materials are in common use for high speed diesel engine pistons, a eutectic silicon—aluminium alloy containing approximately 11% silicon, and a hypereutectic silicon—aluminium alloy with approximately 20% silicon. The latter alloy is harder and provides better resistance to ring groove wear but its thermal conductivity is slightly inferior. Top ring carriers, either bonded cast-iron or a steel rail in the ring groove, are typically used for improved durability with the low silicon alloy but the improved wear resistance of the high silicon alloy may render an insert unnecessary.

When running an aluminium piston in an iron cylinder the greater coefficient of thermal expansion of aluminium can present problems in maintaining small piston skirt to bore clearances under all operating conditions. In order to minimize these clearances and reduce piston slap noise in a cold engine, pistons with expansion controlled skirts are now widely used in passenger car diesel engines. The objective is to limit the thermal expansion of the guiding portion of the piston, the skirt, along the thrust axis of the piston, and this is achieved by casting into the skirt steel struts or hoops.

The function of the piston rings is to seal against the gases from the combustion chamber and against the oil from the crankcase. An indication of how successfully they perform this function is seen in the measurement of engine blowby and oil consumption. To reduce piston height, weight, and friction the sealing requirements must be achieved with the minimum number of rings per piston, and on high speed diesel engines the ring pack usually consists of three rings, two compression and one oil control. To provide good gas and oil sealing the piston rings must possess good strength, wear resistance, scuffing resistance, conformability to the cylinder bore, and freedom from sticking.

A typical ring pack specification is as follows:

(a) Top compression ring of rectangular section with barrelled face either chrome plated or molybdenum spray coated to improve wear and scuffing resistance. Ring width 2 mm but with a trend towards reducing this to about 1.5 mm.

(b) Second compression ring of rectangular section with face tapered back at the top by about 1°. An optional internal step which causes the ring to twist when installed bringing the lower edge into contact with the cylinder wall can improve oil control owing to the scraping action. Ring width 2 mm tending towards 1.5 mm.

(c) Oil control ring of the conformable slotted type with bevelled-edge chrome plated lands. Ring width 4 mm tending towards 3 mm.

In cases where the piston temperatures are high enough to cause top ring sticking problems, for example in a turbocharged engine, then a single sided taper top ring can be specified, the taper angle usually being 7½°. Radial movement of the ring in its groove then alters the vertical clearance, thus minimizing the build-up of combustion deposits and the resulting sticking problems.

Typical materials used for piston rings are malleable cast-iron, spheroidal graphite cast-iron and sintered iron. Some oil control rings with separate top and bottom rails and expander spring are all-steel construction.

The operating temperature of the piston rings is dependent on the cooling of the upper part of the cylinder and on the heat flow into the piston due to combustion. A substantial portion of this heat flow passes out from the crown through the rings and if the top ring groove temperature becomes excessive, that is above about 240°C, ring sticking can occur. In such cases oil cooling of the piston is recommended by directing a jet at the underside of the crown. This oil jet may be fed directly from the connecting rod big-end bearing or the upper half of the main bearing, but neither is recommended because of its adverse effects on bearing oil films and the leakage of pressurized oil from the bearings during cold starting. It is preferable to specify separate cooling jets from an independent pressurized oil gallery with a check valve which does not allow any flow until initial oil pressure has built up in the engine bearings and valve train.

4.5 Cylinder Head
The design of the cylinder head cannot be considered in isolation from the method of valve gear operation and camshaft location which both influence the head layout. Pushrod and rocker activation of the valves from a side mounted

camshaft in the cylinder block is gradually being replaced by direct attack overhead camshaft operation. The lighter weight and greater rigidity of the latter permit higher engine speeds than the pushrod arrangement before problems of valve train malfunction are experienced, and even though, as previously discussed, the practical maximum rated speed for a passenger car diesel engine is probably around 5000 rev/min, this is sufficiently high for overhead camshaft operation to show benefits in valve train dynamic behaviour. In addition the deletion of pushrods along one side of the cylinder head allows more freedom in the design of inlet and exhaust ports and the installation of injectors and heater plugs. Marketing and manufacturing requirements may also dictate an overhead camshaft design if the diesel engine is based on a gasoline engine with this layout.

For passenger car size diesel engines, that is up to about 100 mm cylinder bore, the cylinder head construction invariably used is a one-piece casting covering all cylinders. The alternative constructions of unit heads or paired heads offer some advantages of improved gasket sealing, lower thermal stresses, ease of manufacture and adaptability to families of engines, but the increased weight and cylinder spacing makes them more suitable for larger bore highly rated engines. The one-piece cylinder head offers the most design freedom, the lowest weight, least cylinder spacing and increases engine rigidity. It is also essential if overhead camshaft valve gear is used.

Within the cylinder head there are a number of possible layouts depending on the type of valve gear, the chosen combustion system, the layout of the ports, and the number and pattern of the head retaining bolts. The need to maintain the minimum possible volume above the piston and avoid piston clearance cut-outs makes it desirable to use vertical valves, and space considerations within a small cylinder bore restrict the design to two valves located on or near the cylinder centreline. With pushrod and rocker operated valves there are restrictions due to the pushrod holes on the position of the ports and injectors, and also the combustion chamber on an indirect injection diesel. Unfortunately casting metal thicknesses and minimum water jacket sections do not reduce in proportion to cylinder bore size and a significant problem in designing small diesel cylinder heads is to avoid the merging together of bosses for head bolts, injectors, heater plugs, combustion chambers and valve guides, with the port walls to restrict coolant flow and cause potential thermal problems.

The inlet and exhaust ports may be arranged along one side of the cylinder head, in which case an alternating sequence inlet—exhaust—inlet, etc. is preferable to avoid the thermal problems of two adjacent exhaust valves. Cross-flow porting enables the inlet and exhaust manifolds to be on opposite sides of the head which minimizes the risk of charge air heating and resulting loss of volumetric efficiency by radiation from an adjacent hot exhaust manifold. This layout can also be beneficial if the engine is installed tilted from the vertical because the exhaust manifold can be placed on the underside with the inlet uppermost.

Careful consideration must be given to the number and position of cylinder head retaining bolts if good gasket sealing is to be achieved. The number of bolts per cylinder, some of which will be shared with an adjacent cylinder, may vary from four to seven. Four bolts are usually only satisfactory on the smallest bore engines (below about 85 mm) and stiffening ribs in the bottom deck of the head between the bolt bosses are desirable. Six bolts evenly spaced hexagonally provide a good solution for a direct injection engine but make the installation of a pre-combustion chamber difficult. The latter is better suited by a seven-bolt pattern. In all cases, stiffening ribs between the bolt bosses are desirable to stiffen the bottom deck and improve gasket sealing, and the minimization of drilled and cored holes through the bottom deck also assists this. The cylinder head bolt load per cylinder should be at least four times the applied maximum cylinder pressure gas force, and the bolts are usually tightened to 70% of their proof load. Some typical head layouts are shown in Fig. 9.

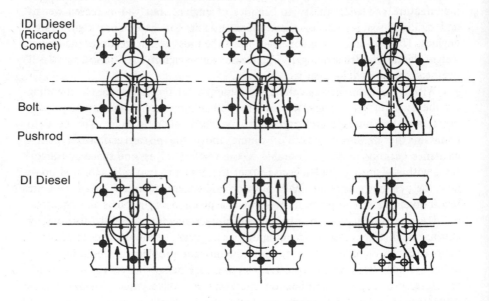

Fig. 9 – Typical cylinder head layouts.

Cooling of the cylinder head is critical and organized coolant flow, preferably from end to end, avoiding stagnant areas and steam pockets is required. The indirect injection system, of which the Ricardo Comet swirl chamber is the most widely used design, is subjected to higher local thermal loadings than the direct injection systems because the jet of burning gases from the swirl chamber results in high heat fluxes in the central area of the cylinder head, particularly the narrow land of metal between the valves. The detail design of this valve bridge region is critical and the metal thickness between gas face and coolant must be reduced, usually by drilling a hole between the valves. Typically, 30% of the coolant flow in the head passes through these valve bridge drillings. This feature is shown in Fig. 10.

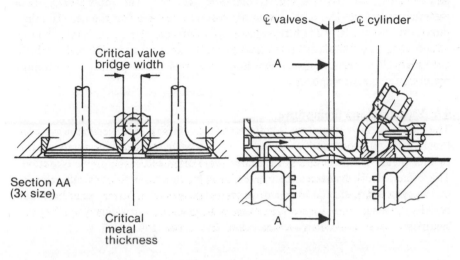

Fig. 10 – Typical Ricardo Comet IDI diesel valve bridge.

The requirements of a cylinder head material are high strength, high thermal conductivity and low thermal expansion. Cast-iron offers a good compromise but aluminium offers the increasingly important advantage of lighter weight. The higher thermal conductivity of aluminium is also beneficial, and its disadvantages of lack of hardness, lower strength and higher thermal expansion are not usually serious in passenger car diesel engines. Magnesium alloy offers a further weight advantage but also suffers from low strength, corrosion problems and high cost. Whichever material is used it is essential to design for ease of casting since high volume production is needed for economy. The water jacket should be a one-piece core, with the inlet and exhaust port cores mounted on sticks for automated assembly into the jacket core. As well as reducing costs such a design ensures the integrity of the water jacket, eliminating mismatch and metal flashing across joints, and also the accurate location of all port cores relative to each other and to each cylinder centreline. This is extremely important with the helical swirl producing inlet ports used in direct injection diesel engines.

Separate valve seat inserts and valve guides are necessary in aluminium and magnesium beads and optional in iron heads, where they provide the opportunity to specify a better material than ordinary cast-iron. The traditional alloy cast-irons are now being replaced by sintered iron alloys and sintered steels with resultant savings in machining costs.

In an indirect injection diesel, such as the Ricardo Comet swirl chamber design, the combustion chamber insert containing the transfer passage or throat between the chamber and the cylinder is subjected to severe thermal stresses. A material having a good resistance to burning, fatigue, and thermal shock, in addition to good hot strength, is therefore required. The most widely used materials are nickel alloys cast using the lost-wax process, but the need to minimize costs, particularly for high volume light duty diesel engines, has led to the adoption of modified casting techniques and cheaper materials such as high chromium alloy steels. A significant alternative development is the use of ceramic materials for these components.

4.6 Valve train and timing drive

The choice of valve operation is between a side camshaft mounted in the cylinder block with pushrods and rockers, or an overhead camshaft mounted on the cylinder head. The advantages of the latter were discussed in the previous section. The principal disadvantages are increased engine height, more difficult removal of the cylinder head, and longer drive from the crankshaft to the camshaft, but for the passenger car diesel engine these are generally outweighed by the very significant advantages of an overhead camshaft layout (Fig. 11).

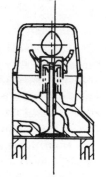

Pushrod and rocker   Overhead camshaft   Overhead camshaft
                            direct acting   pivoted follower

Fig. 11 – Typical valve train layouts.

The performance target for the engine largely determines the valve timings selected. In a diesel engine the inlet valve opening period has to be shorter than in a gasoline engine because the valve cannot be opened earlier than about 12° B.T.D.C. because of the close approach of the piston crown to the valve head, and it should not close later than about 42° A.B.D.C. otherwise engine low speed torque and starting at low ambient temperatures will be adversely affected.

Exhaust valve closing similarly cannot be later than about 12° A.T.D.C. unless valve recesses are put in the piston crown, and this is undesirable because it distorts the combustion space and increases the 'dead' volume in the cylinder at the expense of the combustion chamber volume. Exhaust valve opening is typically about 55° B.B.D.C. The short inlet opening period means that to achieve adequate valve lift in the time available requires high valve accelerations with resulting high inertia forces, and these dictate that valve train stiffness must be correspondingly high. For a pushrod and rocker design the pushrod should be one-piece forged steel with hardened cup and ball ends, and the rocker arm either forged steel or spheroidal graphite (S.G.) cast-iron with hardened pad. The hardenability limit with S.G. cast-iron of about 50 Rc may be insufficient to prevent pad wear and a separate hardened steel pad insert may be required. The pressed steel 'boat type' rocker arm common on gasoline engines may not be sufficiently stiff for a diesel application. The maximum recommended rocker arm ratio is 1.5:1, as higher ratios cause loss of rigidity and also greater magnification of the effect of errors in cam manufacture. Detail design of the rocker shaft and support pedestals for maximum stiffness is important, and overhung end rocker arms are to be avoided. A direct acting overhead camshaft design of course avoids all these problem areas. An overhead camshaft acting via a pivoted cam follower incurs some stiffness penalty and also possibly problems of cam and follower wear due to poor lubrication resulting from the difficulty of providing an oil bath for the cam to dip into during the critical cold-start period before pressurized oil reaches the camshaft and follower.

Pushrod and rocker valve operation requires more frequent valve clearance adjustment than overhead camshaft designs because of the greater number of contact points subject to wear. However, valve adjustment, when needed, on a direct acting overhead camshaft design can be tedious, whereas a pivoted follower can easily incorporate a simple means of adjustment. The use of hydraulic tappets is becoming increasingly common to overcome valve clearance adjustment problems, although there is a penalty in respect of stiffness.

The selection of cam and tappet materials is of great importance to achieve good durability. The most common choice is chill cast-iron for the camshaft and tappet, the latter sometimes being hardened and tempered for improved life, with phosphate coating. A new development in camshaft materials is sintered cams and bearings on a tubular steel shaft, giving significant weight and machining cost savings. A suitable inlet valve material having good strength and wear properties is 1.5% nickel/chromium/molybdenum steel, and for exhaust valves a material with good hot strength such as 21% chromium/manganese/ nickel steel is specified. Chrome flashing of the valve stem is recommended to prevent scuffing.

The choice of timing drive for a side camshaft, pushrod and rocker design is invariably a roller chain because of the short centre distance. For ultimate durability a gear drive may be considered at extra cost. For an overhead camshaft design the choice is between roller chain and toothed belt. The latter offers a significant weight saving and noise advantage but the external belt pulleys require extra oil seals and the belt cover must be effective in preventing ingress of grit, and drifting snow which could freeze on the pulleys overnight. Toothed belt life has been considerably increased in recent years through improved tooth design and materials, and is now comparable to that of a roller chain. As well as providing the timing drive, the same toothed belt can also be used to drive the fuel injection pump and other engine accessories, such as water and oil pumps. However, care is needed to ensure sufficient wrap on the pulleys and correct tensioning, and excessive belt length will increase the effect that stretch and wear have on the camshaft and fuel injection pump timing.

4.7 Fuel Injection Equipment

The indirect injection diesel engine which is used universally in current passenger car applications is invariably fitted with a rotary type fuel injection pump which is significantly cheaper than a corresponding in-line type pump and also incorporates a speed advance of start of injection timing which is essential for the operating engine speed range required. The fuel injector typically used is a screw-in type with variable orifice pintle nozzle which can provide a good performance and noise compromise over a wide speed and load range. The size of these injectors can, however, sometimes cause installation problems in small engines and a new design of small screw-in injector especially suitable for passenger car diesel engines and having a 14 mm thread in place of the usual 24 mm is gaining in popularity. It is important that the high pressure fuel pipes between pump and injectors are as short as possible to minimize the effects of elasticity of the pipe and fuel contents on the injection characteristics. Fuel pipe lengths must also be equal to each cylinder. The position of the fuel injection pump mounting on the engine is therefore of great importance, and it should be as high as possible on the same side as the injectors, or in the centre of a vee configuration engine.

The fuel injection pressures required in the 'wall-wetting' direct injection diesel engine are similar to those in the indirect injection engine and hence a rotary type fuel injection pump of similar specification may be used. In addition a screw-in injector with pintle nozzle is used, but it has a dual-rate injection feature and an extended length nozzle to suit its installation down through the cylinder head.

The higher fuel injection pressures required in the 'toroidal' direct injection diesel engine and the problem of matching the injection equipment over a wide engine speed range are beyond the capabilities of the conventional rotary injection pump. An advanced rotary pump with sufficient pressure capability and electronic control would be required. The in-line injection pump would also require more sophisticated control systems and its cost would be unattractive compared to the rotary pump. The higher injection pressures required make it essential to minimize the working volume between the pump and the nozzle, and the combined pump/injector or unit injector is one method of achieving this. It ensures a much more accurate relationship between cam profile and delivered fuel, but considerations of installation and economics make this approach unattractive for conventional mass-produced engines. The injectors typically used with a conventional pump system are available with various body lengths and diameters but on small engines the limited cylinder head space available can make the 9.5 mm diameter 'pencil' nozzle an attractive proposition. Four-hole nozzles are generally chosen but the hole size, cone angle and nozzle protrusion below the gas face are all critical and are optimized during the engine development.

A cold starting aid is essential for all types of diesel engine and this is invariably an electric heater plug in an indirect injection engine. Various types of rapid warm-up plugs are available and all have automatic control to prevent overloading. Similar plugs may be used in direct injection engines, but alternative starting aids are ether-based fluids or manifold heaters, neither of which carry the performance penalty of the heater plug in the combustion chamber. Heater plugs will enable cold starts to be made down to about $-25°C$ ambient temperature.

4.8 Coolant System

In an indirect injection diesel engine the total heat rejection to the coolant is approximately equal to the engine power output at maximum power, and slightly less for a direct injection diesel. The rate of coolant flow must be sufficient to dissipate this heat without an excessive temperature rise in the system and must ensure an adequate velocity in the various coolant passages, avoiding stagnant regions and vapour pockets. Resultant typical coolant flow rates are from about 60 litre/hp.h for a direct injection engine to about 100 litre/hp.h for an indirect injection, and the minimum required coolant velocity is about 2 m/s. Typically, the coolant pump feeds into the front of the cylinder block with the flow passing along the block and across between the cylinders, and then up into the cylinder head, with adequate flow through the critical valve bridge region, and out from the front of the head. The coolant transfer holes from block to head are sized to give good distribution and to promote flow from front to rear and across the block, and from rear to front and through the valve bridge in the cylinder head. It is preferable to use drilled transfer holes rather than cored holes to maintain maximum rigidity of the block top deck and head bottom deck to ensure good cylinder head gasket sealing. However, this feature is contrary to the requirements for a simple cylinder block casting where an open top deck construction is preferred, and hence a compromise must be reached. The coolant outlet from the front of the cylinder head should discharge into a thermostat to ensure good temperature control and good performance of the vehicle heating system.

The high thermal loading of the central region of the cylinder head, and in particular the valve bridge, on an indirect injection diesel engine has been mentioned previously. As well as providing a drilled coolant hole through the valve bridge it is also important to ensure that cylinder head castings are of good quality and that the coolant passages are clear of sand and flash, particularly around the ports and combustion chamber.

The coolant pump pressure head must be sufficient to overcome external frictional losses due to the radiator and pipework as well as those inherent in the engine. A typical impellor tip velocity is 14–15 m/s at the engine maximum power rated speed.

4.9 Lubrication System
The design of the lubrication system commences with an assessment of oil flow quantity required by the engine, but unfortunately this is not easy to estimate as it depends on a number of design and operating variables such as bearing clearances and oil viscosity. The oil pump must in any case be able to deliver well in excess of bearing requirements at high speed to ensure adequate pressure at

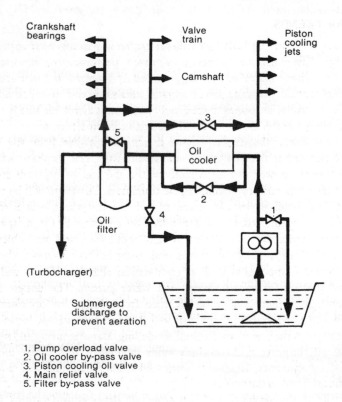

1. Pump overload valve
2. Oil cooler by-pass valve
3. Piston cooling oil valve
4. Main relief valve
5. Filter by-pass valve

Fig. 12 – Typical lubricating oil circuit.

low speed, and also cater for supply to a turbocharger and piston cooling jets if fitted. As an engine wears the oil demand gradually rises, the extent being dependent on the quality of manufacture and the degree of oil and air filtration. However, a pump output of about 25 litre/hp.h at maximum speed is a realistic design starting point.

A typical lubrication circuit diagram is shown in Fig. 12. It is important to ensure that the design oil pressure is attained in the critical areas such as bearings, and this must be achieved as quickly as possible after starting. Hence oil filters and coolers should not drain. A pressure relief valve is required situated near the pump on the delivery side and care must be taken in returning the discharge to minimize aeration of the oil in the sump. A typical pressure relief valve setting is about 5 kg/cm^2.

The heat input to the lubricating oil increases with engine rating, and the growing trend towards pressure charging and the resultant need for piston cooling will therefore lead to increased use of oil coolers to limit the sump oil temperature to about 120–130°C.

5. FUTURE TRENDS

The demands made of the high speed diesel engine in passenger car applications are exacting. The customer expects improved performance, economy and reliability, the vehicle manufacturer requires reduced weight and manufacturing costs, and legislation demands lower exhaust emissions and noise. In addition, engine package size is becoming increasingly important as vehicle size is reduced and body styling dictates the confines of the engine compartment.

It is most likely that future engine designs will evolve from the best of current practice rather than be largely unconventional. The emphasis will be on improving specific power output, mainly withe the use of pressure charging, reducing noise levels and improving engine roughness. The trend will be towards direct injection diesel engines in place of indirect injection, with a resulting economy improvement, but this is dependent on advances in the design of the fuel injection equipment. Significant weight reductions will be achieved by increasing use of light alloys for structural components, ceramics for highly thermally loaded components such as combustion chamber inserts and piston crowns, and plastics for covers, sumps and water pumps. The design of components for light weight cannot be divorced from design for optimum stress distribution and noise emission, and this will demand increasingly sophisticated analysis techniques such as finite element modelling. Manufacturing and materials technology will improve with resulting gains in quality, uniformity and compatibility of components, leading to improved engine reliability and durability with the minimum of maintenance.

The task of the designer in optimizing all these considerations is both daunting and challenging.

3

The passenger car diesel engine - present and future

W. M. Scott, Ricardo Consulting Engineers plc

1. INTRODUCTION

As was predicted some eight to ten years ago, as a result of the fuel crisis, the automotive industry is now in the middle of the 'diesel explosion'. In all the industrialized nations of the Western world, and including Japan, diesel engine passenger cars are selling like the proverbial 'hot cakes', apparently with little regard to the true economics of operating diesel as against gasoline vehicles. Nowhere else has this situation developed more dramatically than in the U.S.A. where the cost of private motoring must be, overall, a lower proportion of the total cost of living than in any other part of the world.

The first examples of high speed compression ignition engines being offered for sale were produced more or less simultaneously by Daimler–Benz and Citroen, the latter using a former mark of The Ricardo Comet Swirl Chamber; later versions of this combustion system, notably the Mk Vb, are now used by practically all the light duty diesel manufacturers. The growth of the passenger car diesel industry within Europe was undoubtedly encouraged by the availability, sometimes illegally, of cheaper diesel fuel which was sold for farming and other purposes free of tax. Consequently when fuel prices began to rise, however justified, the market for diesel passenger cars and light duty vehicles increased until, today, practically every automobile manufacturer feels compelled to offer diesel alternatives, particularly if they seek to enter or remain in the American market.

Ironically, it could be just this same American situation which could put a sudden end to the light duty diesel by the severity of its exhaust emissions legislation. However, it seems unlikely that the most fuel efficient prime mover will be so lightly cast aside. It is the good fuel economy which is the main, if not the only, real selling point for the diesel, although there have been many attempts by technically knowledgeable writers to pass it off as a myth.

The main purpose of this chapter is to present the light duty diesel in its true light and to review the advantages and disadvantages relative to its main competitor, the gasoline engine. Progress in the further development of light duty diesels will also be reviewed.

2. THE PROS AND CONS OF THE LIGHT DUTY DIESEL

If we list the characteristics, both good and bad, of the passenger car diesel engine, in comparison with the gasoline engine, we see that in number the cons outweigh the pros.

    Pros: Fuel Economy
              Gaseous Emissions
              Drivability

    Cons: Prime Cost
              Noise
              Smoke and Particulates
              Engine Weight and Size
              Roughness
              Cold Starting
              Odour

This list presents convenient headings for this chapter which must, of course, commence with the greatest single attribute of fuel economy.

2.1 Fuel Economy

It is now well established that for any one type of power unit, vehicle weight is the most influential parameter and consequently when collecting statistical data they are normally plotted against vehicle weight or inertia, as in Fig. 1. However, since these data have been collected from numerous sources and undoubtedly represent tests carried out under very varied driving conditions, the gasoline and diesel bands are close to overlapping, permitting the observation that some of the best gasoline cars are as good as the inferior diesels. Another point often emphasized by the sceptics is, as is shown, that the diesels are mostly of a lower power/weight ratio and that this is the main reason for the better economy.

By carefully controlled tests on a range of diesel and gasoline cars in which they were all driven at the same average speed over the same town and country circuit it was demonstrated, as is seen in Fig. 2, that the same diesel advantage of 25–30% as in Fig. 1 is achieved. Also of interest in these data are the comparisons at the same power/weight ratio which also conform to the now accepted differential. Typical of such a comparison is that of the 1.6 litre gasoline and 2.1 litre diesels which were fitted to the Peugeot 404 and 504 cars, respectively.

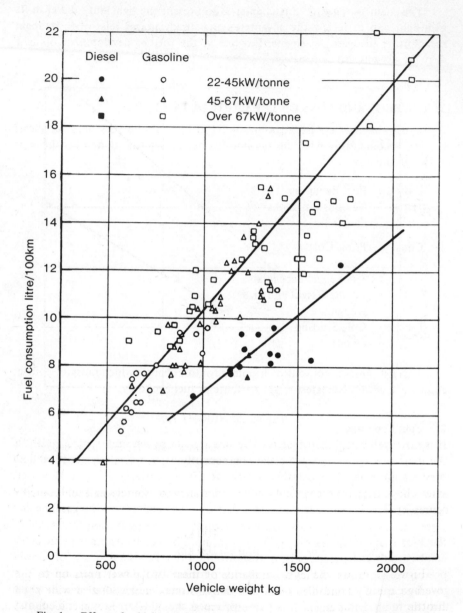

Fig. 1 – Effect of vehicle weight on the fuel consumption of gasoline and diesel cars.

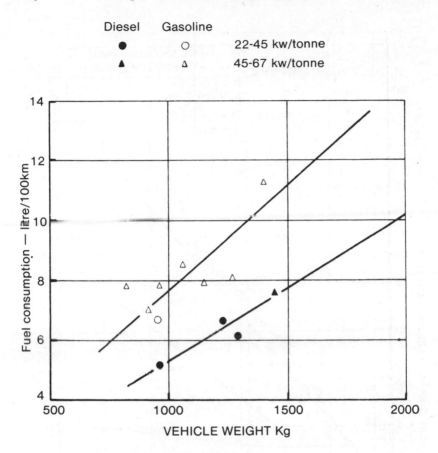

Fig. 2 – Comparison of gasoline and diesel fuel consumptions under closely controlled conditions.

After corrections for the vehicle weight and gear ratios the relative fuel economy figures were:

Gasoline 404   10.45 litre/100 km
Diesel   504   8.30 litre/100 km

Figure 3 shows the close similarity of these two power units up to the governed speed of the diesel and also demonstrates clearly that at wide open throttle the gasoline engine is as fuel efficient as the diesel. It is at part loads and at the lower speeds, where most of us drive most of the time, that the diesel scores. This is clearly seen if we compare the same two engines on a brake specific basis as in Fig. 4. At part loads and low speeds the gasoline engine suffers from the increased pumping losses due to throttling as well as the ever present lower volumetric calorific value of the fuel. At the higher speeds throttling is

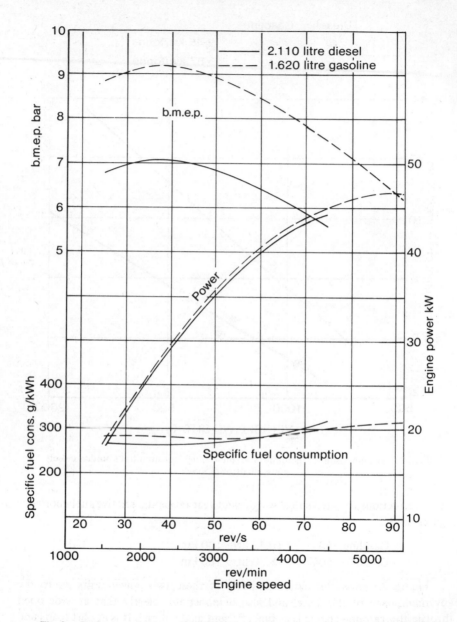

Fig. 3 – Performance curves of comparable gasoline and diesel automotive engines.

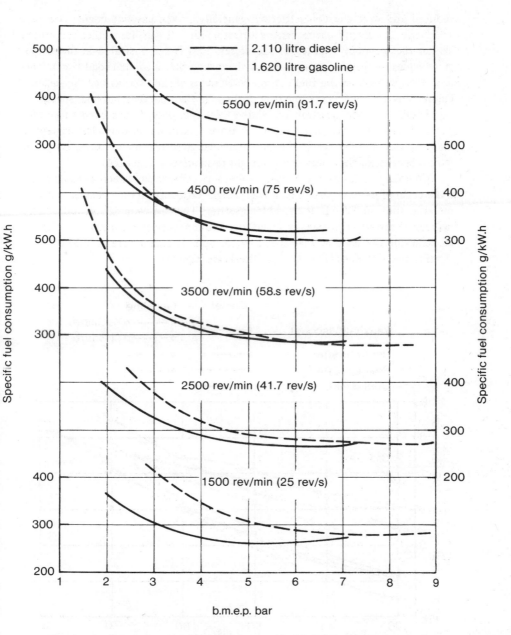

Fig. 4 — Specific fuel consumptions of comparable gasoline and diesel automotive engines.

reduced and with the lower friction, generally, of the gasoline engine even the difference in calorific value can be offset, with the gasoline vehicle returning similar fuel economy figures to the diesel. This is demonstrated in the steady state values of Fig. 5. Figure 6 shows that when the calculated road load curves are superimposed on the fuel consumption maps of the two engines the gasoline engine runs in a more economic part of the map than the diesel at high speeds.

Finally, on the question of relative fuel economy, it has been suggested, perhaps in desperation, that people who drive diesel cars drive in a less aggressive manner because of a need to demonstrate the good reason for their choice, or maybe because of the lower power/weight ratio already discussed.

Comparative tests were carried out by my company between a gasoline and a diesel car of more or less equal power/weight ratio, driven on a chassis dynamometer at the same inertia and road load setting. The cars were driven under three degrees of severity from steady speed to maximum acceleration and braking, but all achieving the same average speed of 15 mph. The cars were a 1600 cc Ford Cortina and a 2100 cc Opel Record Diesel (see Fig. 7).

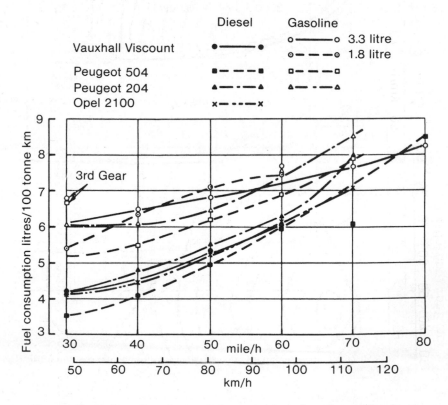

Fig. 5 – Comparison of steady state fuel consumptions of gasoline and diesel cars.

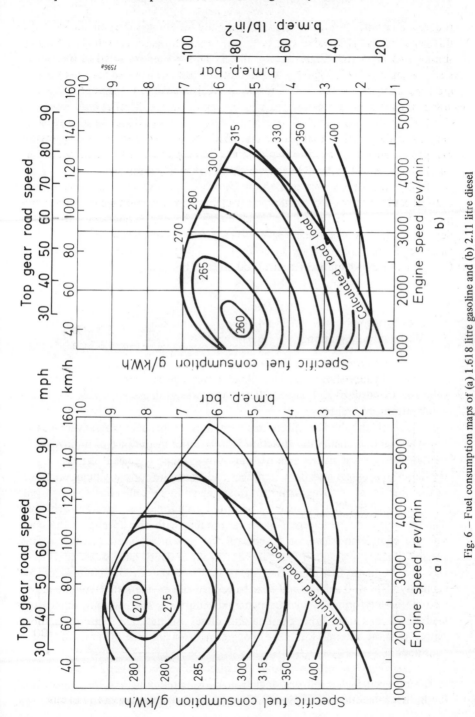

Fig. 6 – Fuel consumption maps of (a) 1.618 litre gasoline and (b) 2.11 litre diesel engines with calculated road load.

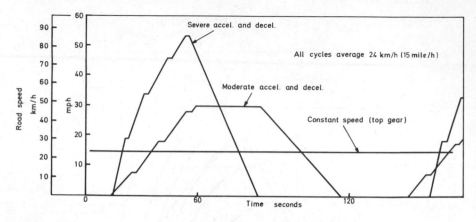

Fig. 7 — Alternative city driving cycles.

The fuel consumptions in litre/100 km and the savings of the diesel car were as follows:

	Diesel	Gasoline	Fuel saving %
Steady 15 mph	4.30	6.81	36.9
Moderate acceleration	6.81	9.44	27.9
Hard acceleration	12.15	18.76	35.2

These figures show that the diesel advantage is maintained however the vehicles are driven and also indicate the wastefulness of excessively hard acceleration and braking.

Although the diesel exhibits this very definite fuel economy advantage over the spark ignited engine the situation is not static; in the face of improvements that are being made in the area of lean burn gasoline engines it is necessary for the diesel to progress also. Two significant advances are currently being employed to this end, i.e. downsizing with turbocharging and direct injection.

Turbocharging has become an accepted technique for uprating light duty diesel engines and has become more acceptable with the advent of the boost pressure controlled wastegate to control the unwanted boost pressure at high speeds. This technique will be dealt with in more depth later when considering ways to improve the power/weight ratio. However, in the context of fuel economy it is well understood that to reduce the engine swept volume and to restore the performance by turbocharging improves the economy under part load conditions from both gasoline and diesel engines. The reason for this is simply that the internal losses are reduced with engine size in terms of the fuel consumed.

It has always been known that the direct injection (DI) combustion system as is employed in larger truck engines has lower pumping and heat losses than the indirect injection (IDI) system. Such systems, employing deep piston bowls,

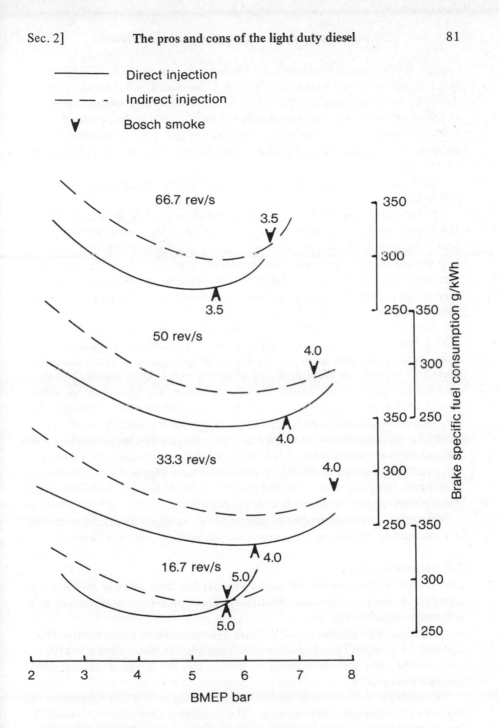

Fig. 8 — Specific fuel consumptions of direct and indirect injection diesels at various speeds.

multiple fuel sprays and induction swirl have been shown to give 10—15% better part load fuel economy as is seen in Fig. 8. These data also demonstrate one of the problems in applying DI technology to wide speed range engines. The mismatch of swirl and injection rate results in loss of smoke limited power at both ends of the speed range. These and the problems of high cylinder pressures, noise and exhaust emissions have until now prevented the adoption of the DI system for passenger cars.

More recent research has gone some way to overcoming some of these problems:

It has been demonstrated that the combustion can be optimized over a wider speed range if injection pressures are increased, thus, allowing the use of smaller injector holes. Although some progress is being made in this direction with traditional pump/pipe/nozzle systems it has also been shown that only with unit injectors can the really high injection pressures be achieved. Figure 9 summarizes the beneficial effect on both smoke and fuel economy of increasing injection pressure.

An alternative approach to direct injection is represented by the MAN 'CDI' system. Based broadly on the original M system it has a single fuel spray directed along the chamber wall as in Fig. 10. However, for light duty applications the single hole injector can be replaced by a variable orifice pintle nozzle. A further modification to limit the needle lift at low speeds and loads enables better control of the spray penetration to match the air swirl. This feature permits the use of lower injection pressures and therefore the employment of rotary pumps similar in specification to those used in IDI systems. Cylinder pressures, noise and emissions are also similar to IDI levels and it is predicted that this system is more likely to succeed in passenger car applications. Again the fuel economy advantage is likely to be 10—15% over the IDI. The main problem is the increased piston height required to accommodate the chamber.

There is a further possibility of improving fuel economy by careful attention to piston and bearing design, valve gear loads, oil viscosity, etc. can all help.

2.2 Gaseous Emissions

The control of emissions from motor vehicles has been steadily increased in severity over the past 20 years. This trend has been pioneered by the U.S.A. and is best represented in Fig. 11.

Until now low gaseous emissions have been counted as a plus for the diesel and still are so when considering untreated engines. However, with a 1.0 g/mile limit on NO_x and the lower limits on particulates the diesel is coming under more serious pressure.

In controlling emissions at source within the engine it is most important to minimize hydrocarbon (HC) emissions. This is because they also form a significant part of the particulate emissions over the light duty cycle and both tend to rise when steps are taken to control or reduce oxides of nitrogen.

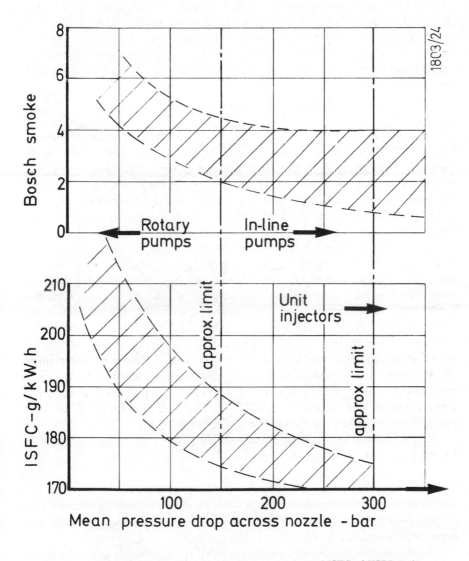

Fig. 9 – Effect of mean nozzle pressure drop on smoke and ISFC of HSDI engines
of 22:1 air/fuel ratio.

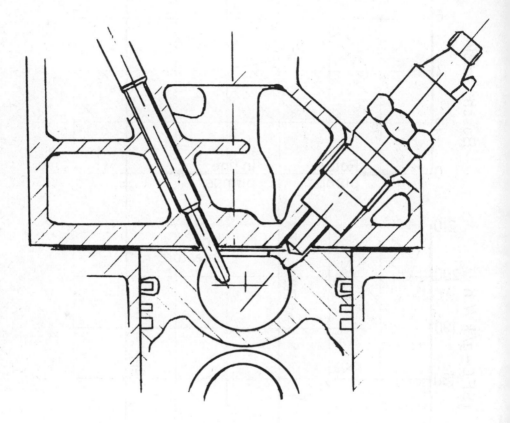

Fig. 10 – MAN combustion system with controlled direct injection and glow plug
in the combustion chamber.

Light duty emissions
U.S. Federal Legislation

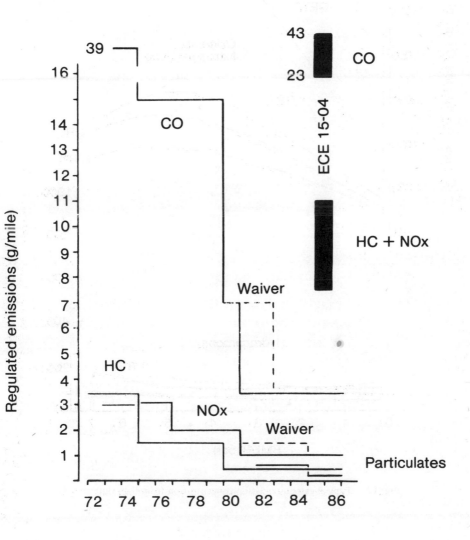

Fig. 11 – History of U.S. emission legislation.

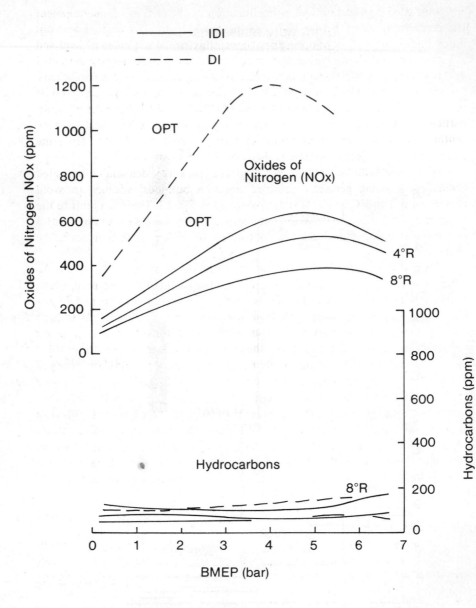

Fig. 12 − Typical gaseous emission concentration from light duty diesels

With good combustion and no malfunction of the fuel injection equipment HC concentrations of below 100 ppm are common for IDI engines and not much higher for well developed DI engines. Only at the extremes of load and injection timing do hydrocarbons rise significantly. If HC levels can be maintained at this level there is no problem in meeting the 0.41 g/mile of the U.S. regulations.

However, the reduction of NO_x by injection timing retard eventually results in increased HC and to meet both 1.0 g/mile NO_x and 0.41 HC becomes difficult, particularly with the larger engines. Figure 12 shows typical NO_x and HC concentrations and the effect of retard of timing. The level of NO_x from the equivalent direct injection system is also shown.

To achieve both acceptable NO_x and HC levels may demand complicated timing plans giving generally retarded injection but local advance to avoid running into high HC levels, as is demonstrated in Fig. 13. There is a limit to the complexity which can be achieved with current hydromechanical timing control systems and only with electronic control are the full potentials likely to be realized.

For larger engines, i.e. over say 2.0 litre total swept volume, NO_x control by timing is not adequate and exhaust gas recirculation (EGR) has to be used. Up to 30% displacement of intake air can be tolerated if confined to an area of the load speed map bounded by 50% load and 30% speed. Fortunately with light duty engines this is the part of the map used most frequently in the emissions test drive cycle. In order to achieve this control both load and speed sensitive signals are required but load signals are not easily obtained from distributor type pumps. Electronic control will help here also.

In addition to HC and NO_x, particulates have to be controlled. Retard of timing can assist with the carbon content since in general for IDI systems smoke improves with retard but not so with DI engines. However, the combined effect on HC of injection retards and the application of EGR both result in increasing particulate emissions. Other than starting with an efficient and therefore clean combustion process the only means of controlling particulates to the low level of 0.2 g/mile will be by after-treatment.

2.3 Drivability
The remaining plus for the diesel is in the area of drivability. The positive control of fuel input under all temperature conditions eliminates the problems of fuel starvation and flat spots, particularly under cold start conditions. Once the engine is running, cold engine power is available on demand without any hesitation.

With the advent of higher power/weight ratios in modern diesel cars it has become necessary to change from the traditional all-speed governor characteristic illustrated in Fig. 14 to a road load governor. Examination of Fig. 14 shows that at any one speed it takes only a very small speed lever or accelerator movement to change from low to high load. This is bad for drivability and for emissions. Road load governors give a characteristic in the middle of the load speed map

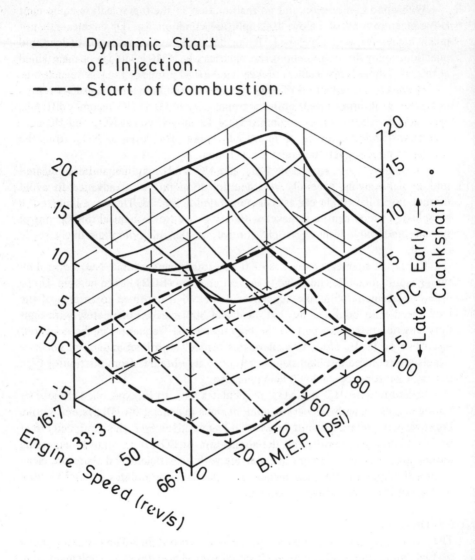

Fig. 13 – Typical timing plan for light duty IDI diesel for low emissions.

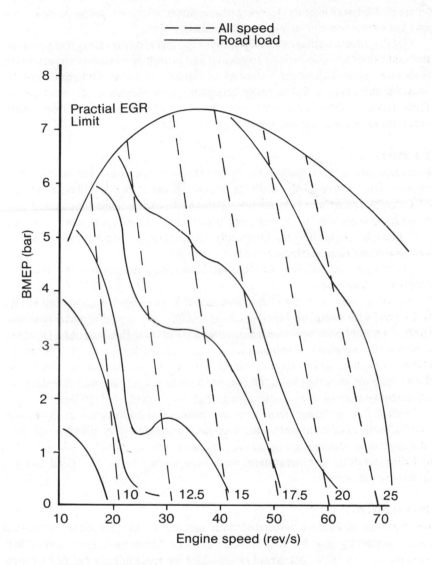

Fig. 14 – Alternative governor characteristics.

where fuel delivery is nearer to being proportional to accelerator pedal movement and this gives a feel similar to a gasoline engine.

Of the three advantages described above the major one by far is fuel economy, the other two being somewhat incidental and not likely to rate as attractions on their own. Against these are a number of critical factors which have to be minimized if the diesel is to be really acceptable as a passenger car power plant. These critical factors may vary in severity and will be dealt with in the reverse order, the less critical factors being considered first.

2.4 Odour

This problem is very subjective in nature and measurements which really represent the degree of the problem are still difficult to make. These measurements are based on the components of the hydrocarbon emissions suspected of being responsible for the characteristic diesel odour. There is, however, no doubt that individuals respond very differently to diesel odour, ranging from unawareness to it being completely unacceptable.

There are two sources of odour which have to be considered – the fuel itself and exhaust odour.

Sniffing samples of diesel fuel and gasoline does not give a good representation of the fuel odour problem. Both are equally unacceptable and if anything, sniffing gasoline gives a greater reaction because of its higher volatility. However, it is this high volatility which also solves the problem. In the case of spillage, for instance, gasoline evaporates so rapidly that the problem quickly disappears. Diesel fuel, on the other hand, persists for a much longer time and therefore it is important to minimize, by good engineering, the possibility of spillage.

Exhaust odour again stems from the incomplete combustion of the heavier hydrocarbons and ultimately can only be controlled by elimination of these components in the exhaust gases. All the means of controlling hydrocarbons, including catalytic after-treatment, have been shown to be beneficial in minimizing exhaust odour.

2.5 Cold Starting

This problem is entirely associated with the waiting time necessary with most forms of starting aid. Starting aids are necessary because, although very high compression ratios are employed c. 20–23:1 in the interests of cold starting, compression temperatures do not reach the level necessary for instant self-ignition until well above temperate ambient temperatures. For instance, without starting aids of some sort, small indirect injection engines will not start on the first fuel injection at normal cranking speed unless the jacket water is at 40–50°C as is illustrated in Fig. 15.

The most popular form of aid is the glow plug, inserted in the swirl or precombustion chamber which when energized electrically to a temperature of about 1000°C acts as an ignition source to promote local ignition.

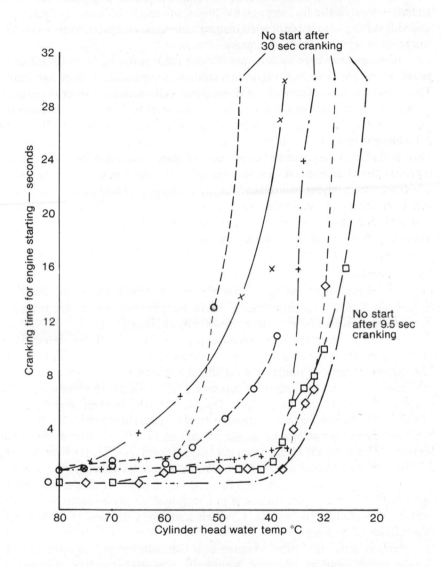

Fig. 15 – Unaided cold starting curves from IDI diesel engines.

Earlier forms of sheathed element plug, designed to project across the chamber towards the fuel spray, took 20–30 s to reach the starting temperature and this waiting time is considered unacceptable when compared with the normal instant starting of the gasoline engine.

Advances have been made towards more rapid warm-up by overloading the plugs electrically and controlling the ultimate temperature within safe limits. This approach has reduced the waiting time, with sheathed element plugs, to 6–8 s and in the case of the Lucas CAV Micronova to 2–4 s which permits the starting aid to be energised simultaneously with the starter motor. This latest development includes a reduction in the thermal inertia of the plug, and as with all the rapid warm-up systems there is an advantage from the emissions viewpoint or even from the point of view of keeping the engine running under very cold conditions in applying continued electrical energy, termed 'post heat', until combustion is sufficiently well established.

Advances have also been made in manifold air heaters and in controlled ether injection but for the passenger car these systems are less preferred than the glow plug.

2.6 Roughness

As with odour, there are two distinct problems which make the diesel less acceptable than the gasoline engine. These are poor secondary vibrations and torque recoil, the latter being the more serious criticism.

For many years the only passenger car diesels were of the four-cylinder in-line configuration which is subject to secondary vibration. Compared with the gasoline engine of similar size the diesel is bound to have worse secondary vibrations owing to the heavier reciprocating parts. Figure 16 compares typical sections of gasoline and diesel piston designs. Note the increase in gudgeon pin diameter which is a major contribution to the increased weight. The degree to which these are felt within the car will depend on the effectiveness of the engine mounts in isolating the engine from the body. Nevertheless, any reduction in the weight of the piston and connecting rod would help in minimizing this problem.

Torque recoil, particularly at idle, is a major problem. Since the diesel is unthrottled the torque fluctuation is not very load sensitive unlike the gasoline engine. This is illustrated in Fig. 17 which also shows the benefit of increasing the number of cylinders.

Various ideas have been considered to eliminate or at least minimize the torque recoil problem, the most technically elegant being that proposed by Heron. Theoretically it is shown that if the engine flywheel is replaced by two wheels, one directly mounted on the crankshaft and a second driven in the reverse direction through gears, then the reaction to the accelerating force on the second wheel, if taken on the engine structure, will cancel out the torque forces causing the recoil, see Fig. 18. The practical problem would be to provide an adequate but silent gear drive that is subject to torque reversals approaching half the engine's instantaneous peak torque levels.

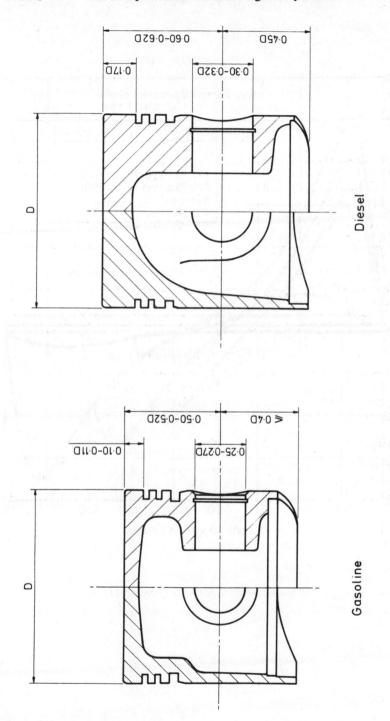

Fig. 16 – Comparison of gasoline and diesel piston designs.

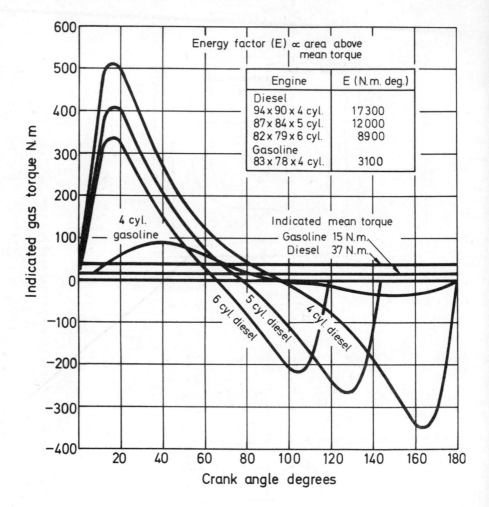

Fig. 17 – Indicated torque diagrams at 10 rev/s, no load.

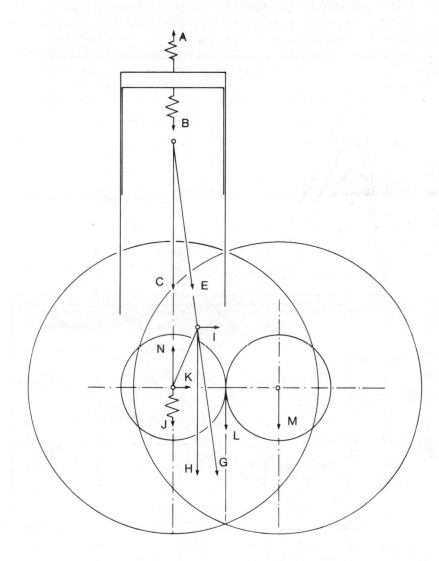

Fig. 18 – Heron anti-recoil system.

In this context it is interesting to review trends in engine configuration. On the one hand the downsizing of engines for fuel economy has resulted in three cyliners being considered as a means of maintaining a reasonable cylinder size and therefore good combustion characteristic. Both secondary balance and torque recoil problems are increased. On the other hand, there are signs of increasing interest in five, six and even eight cylinders where diesels are being considered for upmarket cars. It will be interesting to see how far the market will go in accepting a reduction in comfort in the interest of economy.

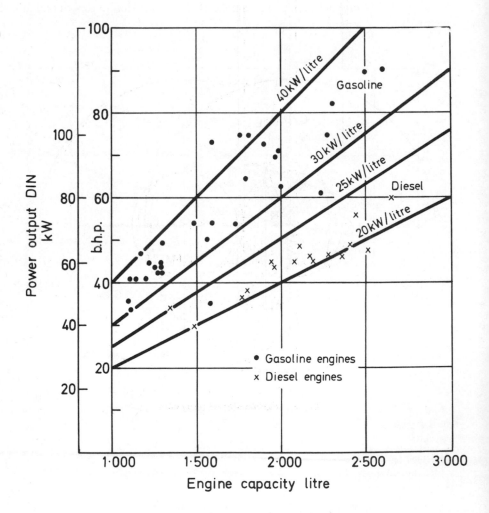

Fig. 19 – Power/litre of some current gasoline and diesel engines.

2.7 Engine Weight and Size

To explain the increased weight and size of the diesel doing the same job as the gasoline engine, it must be remembered that the diesel produces less power per litre of swept volume. This is demonstrated in Fig. 19 and is the combined result of the diesel's lower air utilization and higher friction as is seen in Fig. 20. The lower air utilization is inevitable when considering the fuel air mixing process within the diesel cylinder. No matter what combustion system is employed the percentage of air at TDC presented for mixing is always well below 100, generally nearer 70–75%. The remainder is inaccessible in the necessary working clearances above the piston, etc. The effective mixing of fuel and air is also limited by time and geometry of the fuel sprays and chamber shape, and therefore cannot match the inhalation of a premixed charge.

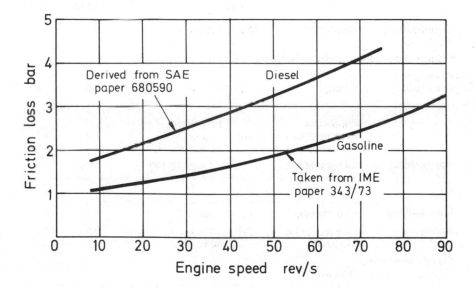

Fig. 20 – Comparative friction losses for 2 litre gasoline and diesel engines.

Comparing the weight per litre of gasoline and diesel engines as in Fig. 21, we see an increase in the region of 12–18%. This increased weight is attributed to the deeper cylinder head to accommodate the combustion chamber and the generally more robust components to withstand the higher cylinder pressures. To these must be added the fuel injection system and the startermotor, both of which are heavier than the equivalent carburettor, ignition system and starter of the gasoline engine. Figure 21 gives a comparison of the major contributors.

An obvious solution to this problem is to employ more light alloy components although such a measure can equally be applied to the gasoline engine.

DIESEL V. PETROL ENGINE WEIGHT ANALYSIS

Component	Diesel Item	Weight kg	Petrol Item	Weight kg	Diesel Penalty kg
Combustion and Ignition equipment	4 cylinder DPA + Delivery Valves	7.13	Distributor Lucas	1.13	
	4 Injectors Complete	1.85	Plugs + HT Lead		
	4 Heater Plugs	.34	Coil Lucas LA12	.79	
			Carburettor	1.47	
		9.32		3.84	5.5
Cylinder Head (Cast Iron)	Statistical	25	Statistical	19	6
Cylinder block difference due to increased height etc.					2
Pistons and Con. Rods	4 Units	8.1	4 Units	5.9	2.2
Crankshaft					0
Flywheel	Insufficient statistical data – estimated only	16.5	Insufficient statistical data – estimated only	12.3	4.2
Starter Motor	1 Lucas M45G	10.5	1 Lucas 2M100	7.9	2.6
					17
Vacuum Pump	1 off Pierburg	1	None	–	1
Battery	1 Lucas BT11A 72 amph	29.5	1 Lucas BT7A 43 amph	19.5	10
Sound Absorption Material	Estimated	2			2
					13
			ESTIMATED TOTAL PENALTY		35.5
Vehicle	Mercedes 220D	1375	Mercedes 220	1335	40
	Peugeot 504D	1279	Peugeot 504	1230	49
	Peugeot 204D XLD	955	Peugeot 204 (1130 cc)	935	20

Fig. 21 – Principal weight differences – diesel and gasoline engines.

There is, of course, a growing interest in pressure charging to increase the power/litre. In addition to ensuring good inlet conditions by taking advantage of induction ram effects there is a choice of three possible pressure charging systems, i.e. turbocharging, pressure wave charging and mechanically driven compression.

Turbocharging — The advent of the very small turbocharger with an integral exhaust wastegate has made possible torque lifts of 40% or so over the upper half of the speed range of a wide speed range diesel without running into excessive cylinder pressures. Because of the starting and emissions limitations mentioned earlier, the compression ratio cannot be relaxed when pressure charging. However, by selecting a turbocharger, matched to the lowest engine speed, it is possible, as shown in Fig. 22, to achieve about 40% increase in torque from 2000 rev/min upwards, falling away to naturally aspirated levels at the lower speeds in conformity with the natural turbocharger characteristic.

Turbocharging in this way naturally increases both the mechanical and thermal loads on the engine and care has to be taken to ensure that the engine design is still viable concerning mechanical loads on bearings and fasteners and metal temperatures in the highly thermally loaded components such as the valve bridge and the piston.

Incorporating an air to air aftercooler assists in all these areas or alternatively will permit more air to be introduced into the cylinder for still higher ratings.

'Comprex' — This is another means of using exhaust energy to generate inlet pressure. The device offered by Brown-Boveri is shown in Fig. 23. The exhaust gases entering one end of the rotor compartments force air out of the other end into the inlet manifold. By suitable porting at each end and by rotating the element, the system can be timed to deliver only air and reject exhaust gas, the rotor being driven from the engine crankshaft. Phasing relative to the engine is not important but the rotor speed has to be about 4—5 times engine speed to match the speed of the pressure waves within the rotor.

The advantage of the system lies in its more rapid response under transient conditions since it does not rely on the speed of moving parts to generate the pressure. Figure 24 compares the performance of naturally aspirated, turbocharged and Comprex boosting of a 2.1 litre light duty diesel engine.

Cost is likely to be the main disadvantage of the Comprex. A mechanical drive has to be provided and the system is more sensitive to exhaust system back pressure. It also benefits from the addition of an air aftercooler in respect of higher rating or reduced thermal and mechanical loads on the engine.

There is a further benefit to be gained from the Comprex system since by modification of the match it is possible to achieve a measure of exhaust gas recirculation. This, of course, can be used to control NO_x emissions, without the need for a separate external system.

Mechanical Superchargers — There is some renewed interest in mechanically driven superchargers of the positive displacement type. The basic problem of

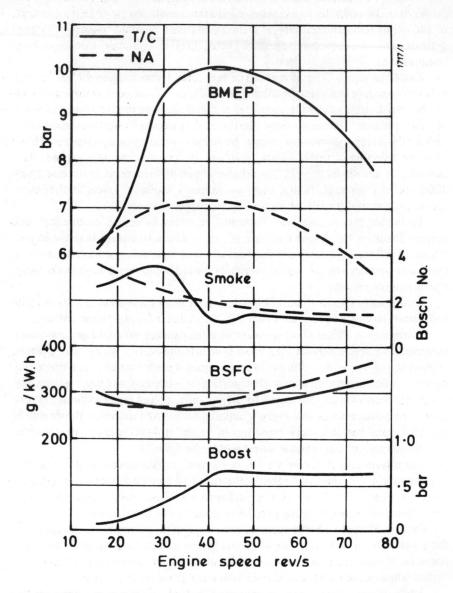

Fig. 22 – Opel Rekord–comparison of turbocharged and naturally aspirated performance.

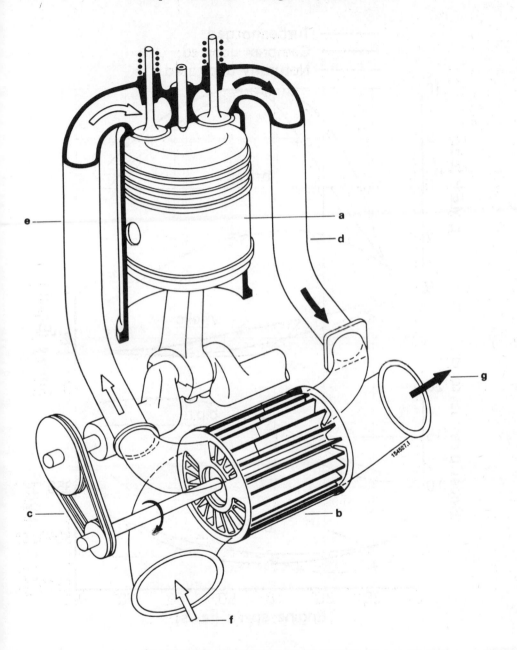

Fig. 23 – Brown-Boveri 'Comprex' system.

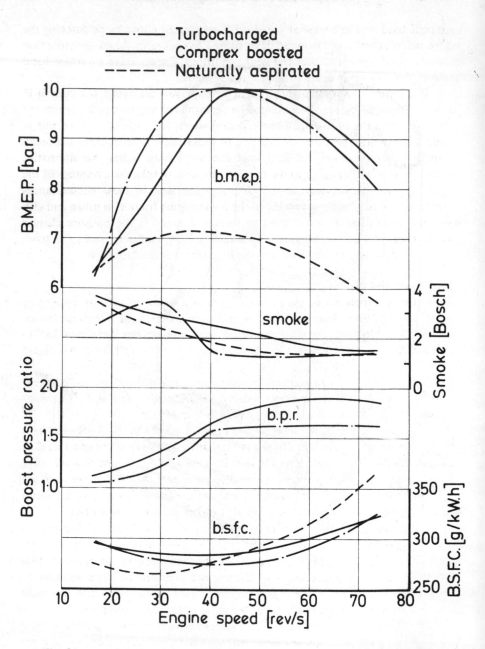

Fig. 24 – Final performance of turbocharged, Comprex boosted and naturally aspirated engines.

parasitic load on the engine at part load is minimized either by declutching the drive when boost is not required or by using a bypass valve. Again the attraction is more rapid response under transient conditions. Experiments are still in hand to complete the comparison of the various systems.

Whichever pressure charging system is employed the result is a benefit in either performance or fuel economy. The performance improvement permits the diesel engine to match its gasoline counterpart on the basis of hp/litre, but at some additional cost and complication. In these circumstances there is usually a small loss of economy at light load due to parasitic losses. An alternative approach which also applies to the gasoline engine is to take the advantage of the higher specific performance to reduce the engine size. In these circumstances part load economy improves owing to the lower engine friction in urban and city driving. The potential for a good performance exists in the pressure charger when required. This technique also benefits NO_x emissions since, in town driving, the boost is seldom used and the throughput is therefore low.

2.8 Smoke and Particulates

These two components in the exhaust gases of a diesel engine rate today as a significant problem. They are basically the same problem since particulates are composed of carbon particles combined with the heavier unburned hydro-carbons, an inevitable product of attempting to burn fuel in the partly liquid state.

Smoke constitutes an aesthetic problem being mainly objected to because it is visible and, in bad cases, causes reduced visibility. Well developed light duty diesels should not present a serious problem in this respect.

However, with air quality control in mind in the U.S.A. the legislators have estimated that the growth in diesel powered automobiles constitutes a threat to the control of particulate material in the city atmosphere and have devised limits which present a serious problem to the diesel engineer. The 0.2 g/mile limit for 1985/6 is a level almost impossible to achieve, even with the smallest diesel engine, without recourse to some form of filtration of the exhaust gases.

Many types of filter have been tried, mostly based on a philosophy of retaining the particulate material within the exhaust system until conditions, i.e. high temperature and excess oxygen, are right for complete combustion. This process is termed regeneration. The filtering and regeneration cycle will have to be automatic and be effective for maybe 50 000 miles. Of course the most efficient collectors also turn out to be the most difficult to regenerate.

As the right condition, i.e. exhaust temperature in excess of 400°C, seldom occurs in city driving some assistance will have to be provided either in the form of a catalyst or a special burner system to heat the exhaust gases. The latter system could run away with a significant part of the saved fuel.

The problem of particulate control is exercising the minds of many people at the present time since if a solution is not found, either the law will have to be relaxed or the diesel will become outlawed in the U.S.A. The latter situation is unlikely to be allowed in view of the growing popularity of diesels in the U.S.A.

2.9 Noise

Probably the main give-away that there is a diesel under the bonnet, especially at idle. Alternatively it has been described as the 'sound of economy'. Nevertheless the reduction of diesel noise receives a great deal of attention and has done so for many years.

Improvements have been demonstrated by controlling the existing forces by smoothing the combustion and by reducing noise radiation by preventing the resulting forces being transmitted to the outer surface of the engine structure. Sometimes, in desperation, enclosures have been built round engines to reduce the noise radiation.

Control of initial injection rate has been demonstrated to be effective in smoothing the cylinder pressure diagram as is shown in Fig. 25. Some means of control exists in the delay action nozzles used in most IDI engines, but further control has generally proved to be non cost-effective. Such control presents further problems with the traditional multi-hole injector DI system although with DI systems able to use pintle nozzles there is clearly a benefit in terms of combustion noise.

Other excitation forces exist to produce mechanical noises often mistaken for combustion knock. Gear driven timing devices are subject to severe impulsive loads from the camshaft and fuel injection pump. These produce noise as the backlash in the gear train is taken up in alternate directions. The advent of the toothed belt has eliminated the problem without introducing the wear problem of chain drives.

Work continues on engine structures to increase attenuation between the combustion and the outer surface. Figures 26 and 27 are typical of these designs incorporating very stiff structures clad with low response outer skins.

Regretfully, although the lessons learned in these exercises contribute to better engine designs, the cost of a diesel as quiet as a gasoline engine under all conditions is still too great to be acceptable.

2.10 Prime Cost and Economics

There is no way that the diesel engine can ever be made as cheaply as the gasoline engine. Compared on an equal basis the engine has a larger swept volume, is heavier and carries some expensive components such as the fuel injection equipment. If the engine has the same swept volume and is supercharged, the situation does not change significantly. The cost of turbocharging is just about equal to increasing the swept volume whereas the alternatives of Comprex or mechanical pressure charger are likely to be more expensive. With the prime cost of the diesel engine being about twice that of the gasoline engine it replaces, it is clear that this fact must be weighed against the saving in fuel costs.

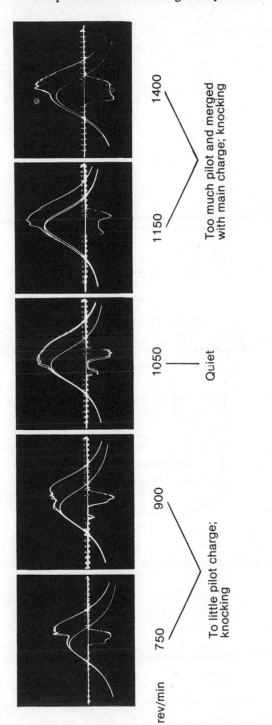

Fig. 25 — Injector diagrams showing effect of unintentional pilot injection.

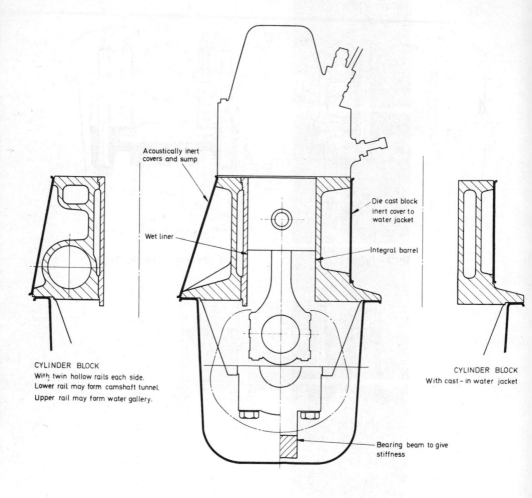

Acoustically inert
covers and sump

Die cast block
inert cover to
water jacket

Wet liner

Integral barrel

CYLINDER BLOCK
With twin hollow rails each side.
Lower rail may form camshaft tunnel.
Upper rail may form water gallery.

CYLINDER BLOCK
With cast-in water jacket

Bearing beam to give
stiffness

Fig. 26 – Schemes for rigid, lightweight, low noise cylinder blocks.

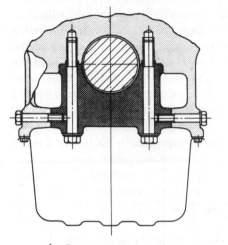

a) Deep skirt type

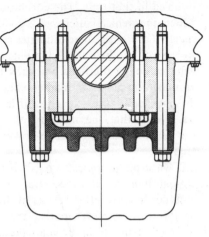

b) Bearing beam type

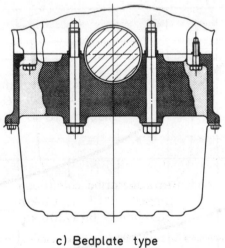

c) Bedplate type

Fig. 27 – V8 crankcase–design alternatives.

Figure 28 represents the trade-off situation as it was about nine years ago assuming that the first cost differential for the vehicle was £100. Diesel fuel at the pump is now about 37.5 pence per litre and vehicle costs have probably also doubled in that time. The conclusion that one has to draw is that the mileage at which the vehicle surcharge is paid for is likely to remain in the 10–20 000 region. Of course, these estimates are based on realistic price differentials related to cost of production of the engines. However, since neither the price of the vehicles nor of the fuels bears any relationship to cost of production and both are based on what the market will stand, the economics of operating a diesel car do not stand close scrutiny. Suffice it to say that a recent management survey has shown that the majority of diesel car owners do benefit financially and the demand for diesel cars is still rising and so consequently is the demand for engines.

On a philosophical note, it is well known that the majority of car owners choose their cars and the extras they want on quite irrational bases and do not consider the cost effectiveness of the additional items. Having paid for the diesel option from whatever motive, the benefit is there for the life of the vehicle in real cash savings or in the enjoyment of a larger car for the same running cost.

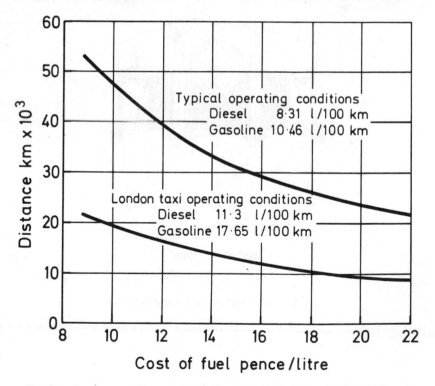

Fig. 28 – Break-even mileage against fuel cost for £100 differential in vehicle cost.

3. CONCLUSIONS

1. The fuel economy advantage of the diesel under all but full throttle operation is undisputed. Although gasoline engines are being improved in this respect so are diesels, with the trend towards direct injection systems.

2. Light duty diesels are coming under heavy pressure in the U.S.A. as regards emissions and although projected gaseous emission limits can be met, the ultimate particulate limit, if not eased, could cause extermination of the light duty diesel in that country.

3. With uprating by the various methods of supercharging, diesel passenger car performance is now more than adequate for the majority of users.

4. Progress is being made towards a more acceptable power unit in the areas of cold starting, noise, vibration, etc.

5. In spite of the dubious economic arguments for the diesel in view of the arbitrary pricing of both vehicles and fuel, the demand is rising and the users of diesel vehicles do benefit significantly.

4

Deformation and stress analysis of engine components using models

H. Fessler, Professor of Experimental Stress Analysis, University of Nottingham

1. INTRODUCTION

Deformations of a number of engine components need to be known so that they may be limited or to allow stiffness to be calculated. Excessive deformation may cause failure of fluid film lubrication or it may cause fractures. Certain relative deformations of components which are pressed together can cause fretting and consequent fatigue failure. Some deformations need to be known to determine stiffness so that natural frequencies may be calculated.

Critical stresses in engine components need to be known to ensure reliable operation. At the initial design stage, 'strength of material' type calculations, which simplify parts to bars, cylinders, flat plates, etc. in tension, compression bending, torsion etc., are used to obtain first approximations to dimensions. If the positions and directions of the forces (and reactions) acting on each component are carefully chosen at this stage, the stresses calculated in this way can be used as the nominal stresses to be multiplied by appropriate stress concentration factors to obtain prediction of maximum stresses for use in fatigue life calculations.

Published stress concentration factors (e.g. [1]) are unlikely to be adequate because they usually apply to simpler shapes and loading modes than found in modern engines. After the shapes have been drawn which would form suitable boundaries of the various fluids and are expected to be able to apply the necessary forces to contain the fluids, stress analysis is required.

Stress analysis can be carried out using photoelastic, numerical (e.g. finite element) or other models. These last require strain gauge measurement and are convenient when engine components are available. These techniques give peak stresses owing to particular, calculated loads acting on a component. If the loads are not known, they can be measured on running engines; in some cases it is possible to locate strain gauges in running engines so that peak stresses at critical points can be recorded for a complete engine cycle.

This chapter concentrates on the use of epoxy resin models to determine deformations and stresses.

2. MANUFACTURE OF MODELS

Models of simple shape (e.g. studs, valves, crankshafts) can be machined from prismatic bars or sheets. Complicated shapes (e.g. cylinder blocks, cylinder heads, sumps, gear casings) are most efficiently produced by precision casting [2, 3]. The dimensional accuracy of the latter is as good as the 'lost wax' process for metal castings. Only contacting surfaces of such models need to be machined.

Araldite is the best available model material because it is chemically stable, easy to cast in the laboratory and has adequate optical sensitivity for photo-elastic analysis. Moulds for simple shapes may be made of any impervious substance which can withstand the pouring temperature ($135°C$) and the hydro-static forces exerted by the liquid resin. They must be dry because water impedes polymerization. Because the material shrinks during curing, it is easier to avoid shrinkage stresses (cured into the casting) if cores are made collapsible. As epoxy resins are good adhesives, all mould surfaces have to be coated with a separating agent; silicone liquids or greases are convenient substances for this purpose. For castings which are to be machined on the outer surfaces, sheet metal containers of simple shape are convenient. If the model has large cavities with machined surfaces, cylindrical cores can be constructed from rubber tubes and rubber sheets.

Precision castings are made using patterns of any dimensionally stable material. Traditional wooden patterns are expensive because the accuracy required is much greater than that for metal castings. Metals and Araldite are often more convenient materials. Moulds and cores are made of mixtures of slate powder and (room temperature curing) epoxy resin.

Models can be machined like metal prototypes but care is needed because the material is brittle (it is a long-chain polymer in its glassy state). Adequate cooling (by liquid or air) is essential because the material is a thermal insulator which can be damaged by additional curing caused by overheating.

3. METHODS TO DETERMINE STIFFNESSES

The simple expression for stiffness, $K = P/\delta$, i.e. force/displacements, needs careful definition. The positions of the reactions as well as the particular load P and its extent and the position and direction of the resultant force must be specified. It is also important to decide the displacement δ of which point on the component relative to which other point should be considered when calculating stiffnesses. This shows that some components have many different important stiffnesses.

The practical methods to determine deformation are.

3.1 Simple 'strength of materials' type calculations

Treating parts of components as beams or plates or cylinders may be considered as dividing the component into large elements. These 'usual engineering calculations' are cheap and quick but often inaccurate because the boundary conditions needed for the calculations are often unrealistic and the beams etc. are often too short for the basic assumptions of the theory to apply (e.g. conventional beam calculations applied to the webs of a modern crankshaft). Furthermore, continuity is usually not satisfied at the junction of these large elements. Sometimes (e.g. [4]) it is possible to apply published theory-of-elasticity solutions to obtain overestimates of measured stresses.

3.2 Finite element calculations

Finite element calculations do not have the technical disadvantages of the above but they are often very expensive owing to the large numbers of elements needed to model engine component shapes.

3.3 Measurements in running engines

Accurate measurement of deformations on running engines is very difficult owing to vibrations, temperature variations and the inaccessibility of most of the critical surfaces. Where it can be done, it is useful because the results are independent of any estimates of loads on components.

3.4 Measurements on engine components in rigs

Measurement on engine components in rigs is often cheap for small, existing engines because components are available or are easily modified from existing parts. For large engines the loads and therefore the loading rigs and their cost can be large.

Most rigs are static and apply only one predominant loading mode. The relation between the magnitudes of the rig loads and the corresponding engine loads is usually complex and sometimes the most difficult part of the whole test.

3.5 Measurements on models

Measurements on models cause the same difficulties as rig tests but models can be made to reduced scale and of different materials from those used in the engine. Using a small scale and low-modulus model materials can reduce large-engine, component tests costs significantly.

The deformations of metal or acrylic (perspex) models have to be measured while the loads are applied (e.g. [5,6]). The deformations of epoxy (Araldite) models may be measured directly like those of other materials, or the frozen deformations produced by the controlled heating and cooling under load used for frozen-stress photoelasticity may be used (e.g. [7,8]). Direct measurement is quicker and more accurate, whereas frozen measurements allow detailed description of deformation of bearings and other contact surfaces because the unloaded (and therefore dismounted) components are measured. However accurate values must be the differences between loaded and unloaded positions of selected points in the surfaces; this requires careful positioning of the model.

4. DEFORMATIONS OF CRANKSHAFTS

At any one instant, each connecting rod exerts a force on the crankshaft and, fortunately, each connecting rod force can be taken in turn and subsequently their effects can be summated (because the system is linear, at least as a first approximation). The oil film between the crankpin and connecting rod exerts pressure on the crankpin. The elastohydrodynamic pressure distribution cannot be determined until the instantaneous position of the connecting rod relative to the crankpin is known, but we would need to know much about the stiffness and all the loads acting on the crankshaft, crankcase and other components before this could be determined. A uniform pressure distribution in the oil film is therefore assumed.

Because real engines have more than two main bearings, the crankshaft is a multi-span beam with, say, four to ten supports; the positions of the reactions at all these supports depend on the bearing clearances, other loads on the crank-shaft and many other factors. To make progress, further simplification is necessary. It is usually assumed that the reactions act in the middle of the bearings.

The next stage is to define the important displacement. If the main bearings are under consideration, the slopes of the journals relative to the bearings are important. Our experience (or simple elementary strength of materials) tells us that, in the plane of the crankthrow of a modern crankshaft, the webs are much more flexible than the journals. This is illustrated in Fig. 1. It is therefore reasonable to use the 'web-spread' as a measure of the slope of one journal relative to the other. An example is shown in Fig. 2.

The total difference in slope of the journals does of course depend on the vector sum of the components in-plane and out-of-plane relative to each particular crankthrow AND the effect of the adjacent crankthrows. It depends on the sophistication of the subsequent analysis whether such precise, detailed information can be used.

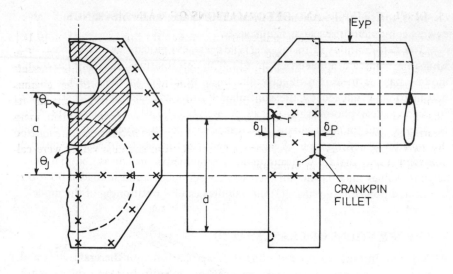

Fig. 1 – Crankshaft model (×indicates distortion measuring position).

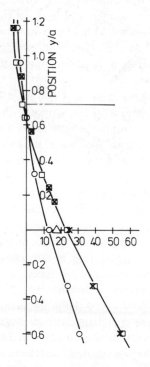

Fig. 2 – Web spreads in two different crankshafts due to radial force on crankpin.

5. DISPLACEMENTS AND DEFORMATIONS OF MAIN BEARINGS

A Vee engine with underslung crankshaft was chosen for this investigation [9,10] since this type of engine is used in competitive, high output applications. The Vee angle chosen was 45° since this is most widely used. Twelve-cylinder models were made with 'side by side' connecting rods because 12-cylinder engines produced high bearing loads, particularly at intermediate bearings. The model shape, shown in Fig. 3, is a series of box-like structures, consisting of plain, main bearing panels connected by plain, longitudinal walls. The bearing caps are located by two vertical studs and two horizontal studs. Stud stiffness values were calculated from prototype information.

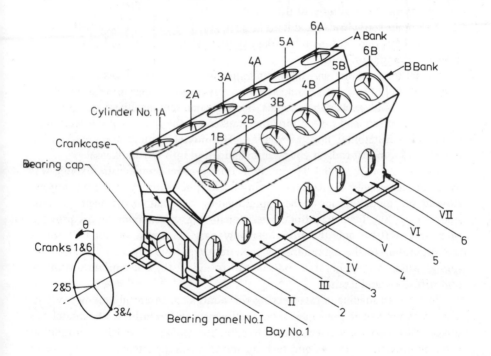

Fig. 3 – Model of V-12 cylinder block. Reproduced from [9].

Because its stiffness greatly influences the magnitude of the loads applied to the crankcase, the crankshaft was included in the loading system. Static model contact stresses cannot model the prototype engine bearing conditions with dissimilar metals, oil film and bearing shells. The last-named were made integral with the crankcase and bearing cap. The main bearing clearance was made 0.07% of the journal diameter.

Gas pressure and reciprocating inertia loads were identified as the most significant dynamic loads which occur in prototype engines. It was decided to load the model with gas pressure and reciprocating inertia forces acting along the cylinder axes. The maximum gas pressure force was obtained from an engine survey; the ratio of maximum gas pressure force to reciprocating inertia force (4:1) and variation of gas pressure force with crank angle were obtained from an engine manufacturer. The variation of reciprocating inertia force with crank angle was calculated from the mean ratio of connecting rod length to stroke length (2.34) obtained in the engine survey. The firing order chosen for the application of gas pressure forces was 1A−3B−3A−5B−5A−6B−6A−4B− 4A−2B−2A−1B since this order produces significant pressures in adjacent bays. Figure 3 shows the positions of the cylinders.

The main static loads are due to bolt (or stud) forces and engine self-weight. The most important are main bearing cap vertical and horizontal bolts, cylinder head and foundation bolts.

Model manufacture and the room temperature loading devices are described fully in [9,11]. The model crankcases were precision castings in Araldite CT200 epoxy resin using expendable moulds. Bearing caps were machined from pre-cast epoxy resin sheets and bolted into the crankcases; the main bearings were line-bored on a horizontal boring machine. The model dummy crankshafts were turned from pre-cast rods of epoxy resin and fitted into the crankcase.

Measurement of model deformations during room-temperature loading was by linear capacitance displacement transducers. The accuracy of the transducers allowed the model strains to be approximately one half of the engine strains. Accordingly, the load magnitudes were small (e.g. the maximum gas pressure force was 440 N) and therefore relatively easy to apply. Model bolt forces (corresponding to the maximum gas pressure force) were applied by conical disc springs which were calibrated and selected to model the appropriate prototype bolt stiffnesses and loads.

For frozen loading, stress-freezing was conducted in an oil tank in a hot air oven; oil was used to prevent excessive self-weight deformations. The model was subjected to two stress-freezing tests to allow the model self-weight deformation to be eliminated. In the second test, the model was supported as in the first and loaded by a system of dead weights and levers as shown in Fig. 4; bolt forces were exerted by helical steel compression springs. Gas pressure forces acting on the cylinder top-deck plate of the appropriate cylinders were applied by levers similar to that applying an upward crankpin force. This loading is for one of the critical crank positions, $\theta = 710°$. The model was measured before and after stress-freezing at 20 points in each main bearing bore and corresponding journal positions and at 11 points around each panel periphery. The absolute radial displacements of each point in the journal and bearing surfaces were obtained from these measurements.

The deformed bearing and journal shapes of the stress-frozen model pre-

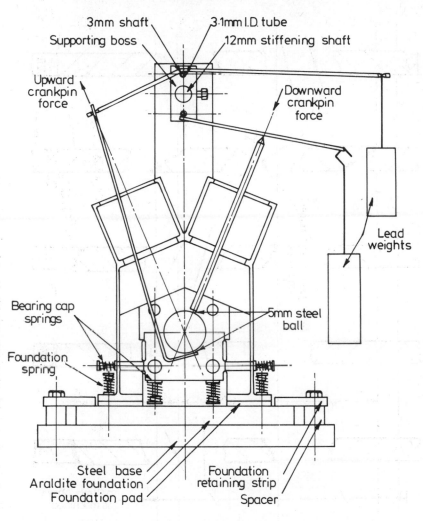

Fig. 4 — Frozen loading system for crank case model. Reproduced from [10].

sented in Fig. 6, for the vertical plane only, also show the journal diameter, the
initial bearing bore diameter and the bearing bore diameter after clamping (i.e.
application of bearing cap bolt forces only). Agreement between the room
temperature and frozen displacement and shapes was good. At the crankshaft
angle of 710°, cylinder 1B is exerting the maximum gas force; the inertia forces
in bays 1 and 6 are in the upward direction, in bays 2 to 5 in the downward
direction. The effect of these forces is to produce overall sagging of the model,
contact between the top of the bearing and journal at bearings VI and VII and a
large vertical journal slope at bearing I. Figure 5 shows how the crankshaft takes

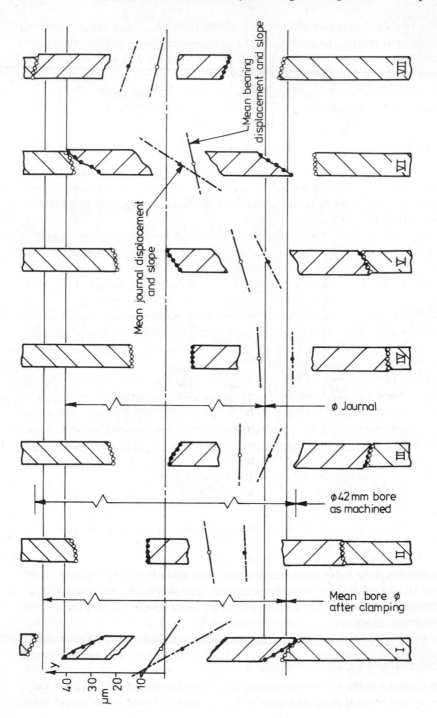

Fig. 5 – Vertical bearing (○) and journal (●) displacements of model V12-1 due to engine cycle loads. Reproduced from [10].

up the 'vertical' clearance in the seven bearings and indicates the accuracy of these measurements. The mean journal displacements and slopes can be used to calculate the stresses set up in the crankshaft due to crankcase deformation. It should be noted that, owing to the bearing clearances, the journal slopes and displacements differ significantly from the mean bearing values. The corresponding horizontal values are small.

6. METHODS OF STRESS ANALYSIS

Stresses in engine components may be calculated using simple strength of materials or finite elements as discussed for stiffnesses. Alternatively, stresses may be obtained from strain gauges in running engines or in rigs or using photoelasticity (e.g. [12–14]).

For both types of strain gauge investigation it is important to note that few important engine parts are under uniaxial stress. It is therefore necessary to use rosette strain gauges (or Tee gauges where the directions of the principal stresses are known owing to symmetry of shape *and* loading) to calculate the surface stresses from the measured strains using the generalized Hooke's law.

Rig and model tests require the definition of loads and their conversion to suitable, usually static values. Because these methods usually test one component only, the proper representation of the restraints exerted by adjacent components in the running engine needs great care.

Photoelasticity may be used with flat models, loaded in their own plane, stresses being determined from the interference effects when viewed in polarized light. This 'direct' method is rarely useful for engines because few parts of engines can be well represented by plane stress models. The only practicable three-dimensional method is the frozen stress technique (e.g. [15]). Any steady load may be applied to models which are heated and cooled under load before slices are cut from the model and analysed photoelastically.

The equivalent steady loads may be applied to frozen-stress models by freely hanging weights, compression or tension springs, pneumatic or hydraulic pressure. The large, non-linear, thermal expansion of epoxy resins (the only practical model materials) and their very low modulus at elevated temperatures must be considered in the design of loading rigs.

Stresses can be determined at any position in the model but interest is usally limited to surface stresses. Most investigations aim to produce elastic stress concentration factors which can be applied for different engine running conditions and to similar components in other engines.

7. CRANKSHAFT FILLET STRESSES

Crankshafts of multi-cylinder engines may be considered as bars of complicated shape on elastic supports, which are mainly subjected to transverse forces. The tangential components of these forces cause bending and torsion, and radial components cause bending only. The torque is transmitted to the end of the shaft as useful output. Because the only highly stressed regions are in the fillets shown in Fig. 1, an efficient method of stress analysis is to study the stresses in a single throw. The loading modes which act on a single throw are shown in Fig. 6. The stresses are determined for as many of these as are considered necessary. A radial force (F_{yP}) acting on the crankpin, exerted by one of two side-by-side connecting rods, is used here to describe the method; further results are given in [16].

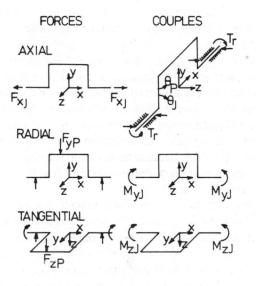

Fig. 6 – Loading modes and sign convention for crankshaft tests. Reproduced from [16].

Models, as shown in Fig. 1, and dummy main bearings and big ends were made of Araldite. The main bearing blocks were simply supported and free to move axially to allow for differential thermal expansion of Araldite and steel. The connecting rod force was applied by a freely hanging weight. After the stress-freezing cycle, deformations were measured to determine web spreads like those shown in Fig. 2. Slices were cut from the model to contain the fillet region in a number of θ = constant planes (defined in Fig. 1). The 1.5 mm thick slices were cut using a diamond-impregnated, mild steel slitting wheel in a high speed milling machine. The slices were analysed in a transmission polariscope [17]. The material was calibrated by determining the Material Fringe Value from rectangular cross-section, tensile specimens cut from annealed parts of the model after the required slices had been obtained.

All stresses are expressed as multiples of nominal stresses, called stress indices I. These nominal stresses at the centre of the crankpin were calculated from the applied forces and the undeformed shape of the models.

Both principal stresses σ_1 and σ_2 in the surface of the crankpin and journal fillets were determined over extensive regions of these toroidal surfaces. An example of the results is shown in Fig. 3. Measurements were taken in $\theta =$ constant planes at a sufficient number of positions to define the curves. In each plane these measurements were taken at various inclinations α (see Fig. 7) but it proved more convenient to show the results in their true axial positions x_P.

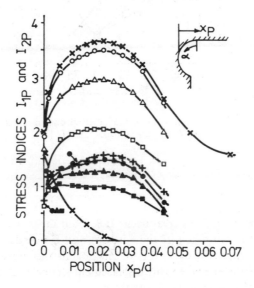

Fig. 7 – Stress distributions due to radial force on crankpin F_{yP}. Reproduced from [16].

Position, $\theta°$	0	15	30	45
Greater stress index I_1	X	o	△	□
Smaller stress index I_2	+	●	▲	■

Models of shapes V, H, KVM and KM were loaded by radial forces on the crankpin (F_{yP}), the loading mode which is generally considered most important. The results for H, KVM and KM were similar in type to those shown in Fig. 7. Because stresses due to F_{yP} are symmetrical about the $z = 0$ plane, the greatest values in the $+\theta$ planes only are shown in Fig. 8. The positions of these greatest values in the crankpin fillet were at $\alpha \simeq 30°$ (defined in Fig. 7). The directions β of the principal stresses are only shown by chain-dotted curves. They differed little between the different shapes; their individual values have been omitted to avoid confusion.

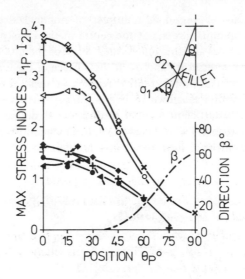

Fig. 8 — Maximum stress indices in each θ position due to radial force on crankpin F_{yp}. Reproduced from [16].

Shape	V	H	KVM	KM
I_1	X	◁	◌	○
I_2	+	◀	◆	●

The peaks of these $I–\theta$ curves occur at $\theta = 0$ if there is no hole through the web near the fillet considered. If the adjacent part is bored, the peak moves to $20° \leqslant \theta \leqslant 25°$ approximately; this occurs for all the journal fillets and the crankpin fillet of the H shape. This displacement of the peak is associated with the reduction of the value in the $\theta = 0$ plane.

Because the directions of the principal stresses due to different loading modes differ, the principal stresses due to several loadings must be determined by vector summation of the values for the individual loads. As the positions of the greatest stresses do not coincide, this summation must be carried out over areas of each fillet. The position (θ, α) as well as the magnitude of the combined principal stresses depends on the values of the nominal stresses and the stress indices due to each loading node.

ACKNOWLEDGEMENTS
The author acknowledges the contributions of his co-authors listed in the references and thanks the technicians of the Department of Mechanical Engineering for their skilled, enthusiastic work over many years. Illustrations are reproduced from papers published by the American Society of Mechanical Engineers, the Verein Deutscher Ingenieure and the Science and Engineering Research Council.

REFERENCES

[1] Peterson, R. E., *Stress Concentration Factors,* 1st Ed., John Wiley & Sons, (1974).

[2] Fessler, H. and Perla, M., 'Precision Casting of Epoxy-resin, Photoelastic Models, *J. Strain Anal.,* **8**, 30 (1973).

[3] Fessler, H., Little, W. J. G. and Wentchead, P. S., 'Precision Casting of Epoxy-resin Models Using Expendable or Reusable Moulds', *8th All Union Conf. Photoelasticity,* Tallinn, U.S.S.R. (1979).

[4] Fessler, H. and Ollerton, E., 'Stresses in Internal Combustion Engine Poppet Valves', *J. Mech. Eng. Sci.,* **6**, 1–8 (1964).

[5] Ahmad Bin Chik and Fessler, H., 'Radial Pressure Exerted by Piston Rings', *J. Strain Anal.,* **1**, 165–171 (1966).

[6] Feslser, H. and Perla, M., 'Photoelasticity Applied to Complicated Diesel Engine Models', Institute of Physics, *Stress Analysis Group Annual Conference,* 17–19 September (1974), pp. 111–122.

[7] Fessler, H. and Sood, V. K., 'Deformations of Some Medium Speed Diesel Engine Crankshafts', Paper No. 23, *CIMAC Conf.,* Washington, U.S.A. (1973).

[8] Fessler, H. and Perla, M., 'Deformation of Diesel Engine Frames from Static Model Tests', Paper No. 73–DGP–1, American Society of mechanical Engineers, *Diesel and Gas Engine Power Conference and Exhibit,* Washington, D. C., 1–4 April (1973).

[9] Fessler, H. and Whitehead, P. S., 'The Use of Precision Cast Models to Predict Diesel Engine Deformations', Paper No. 313, *International Conference of Stress Analysis,* September (1978), pp. 97–106.

[10] Fessler, H. and Whitehead, P. S., 'Deformations and Stresses from Statically Loaded Diesel Engine Models, *Universities' Internal Combustion Engine Group Proceedings,* April (1980), pp. 221–229.

[11] Whitehead, P. S., 'Stiffness and Strength of a V-type Diesel Engine Crankcase', *Ph.D. Thesis,* Nottingham University, 1978.

[12] Fessler, H. and Padgham, H. B., 'A Contribution to the Stress Analysis of Piston Pins', *J. Strain Anal.,* **1**, 422–428 (1966).

[13] Fessler, H. and Perla, M., 'Stresses in a Medium Speed Diesel Engine Frame', The Institution of mechanical Engineers, *Applied Mechanics and Combustion Engines Group Conferences Proceedings* (1976), **190**, 21/76, pp. 309–318.

[14] Diver, P. and Fessler, H., 'Stress Concentrations at Dimples in Crankshafts', *J. Strain Anal.,* **9**, 78–81 (1974).

[15] Durelei, A. G. and Riley, W. F., *Introduction to Photomechanics,* Prentice-Hall, 1965.

[16] Fessler, H. and Sood, V. K., 'Stress Distributions in Some Diesel Engine Crankshafts', Paper No. 71–DGP–1, American Society of Mechanical Engineers, *Meeting of the Diesel Gas and Power Group,* Toronto, Conada, April 18–22 (1971), pp. 1–10.

[17] Fessler, H., 'The Collection of Photoelastic Data', *J. Strain Anal.,* **3**, 128 (1968).

5

Practical applications of finite elements in the stressing of diesel engine components

A. K. Haddock, Perkins Engines

1. INTRODUCTION

Before looking at the tools available to the design analyst today, the historical approach to the design and development process will be discussed. In the past, the designer would have access to classical stress analysis techniques, but these, although useful for approximate sizing of simple components, are totally inadequate for assisting in the understanding of, for example, the temperature distribution in a cylinder head or the stresses in cylinder blocks. This resulted in the traditional approach to the introduction of a new engine. The engine would be designed using classical analysis, built, and the experimental engine would then be rigorously tested. Components which failed would be modified and retested. The whole process was thus one of evolution to the final product and all too frequently the project would be 'development' led.

With the advent of large, high speed computers the engineer now has the tools to allow him to solve the structural and thermal equation systems which describe the nature of complex engine structures. In particular, Finite Element (FE) analysis allows the performance of engine components to be simulated at the design stage.

We will therefore be looking at an approach to engine analysis which provides significantly increased confidence in engine integrity to be established prior to component manufacture and test. Hence significant reductions in high cost of development can be made.

2. OBJECTIVES OF ENGINE FINITE ELEMENT ANALYSIS

The overall objective of conducting engine analysis is to ensure that the total 'integrity' of an engine is at a maximum, prior to costly manufacture of hardware.

The scope of FE analysis goes beyond just stress analysis. It covers component temperature prediction, sealing (e.g. major gasket performance), vibration of crankshafts (TV) and high frequency cylinder block forced vibration for use in noise prediction.

Although not specifically the subject of this chapter, FE analysis of bearing oil films allows detailed bearing design features to be assessed.

The use of FE structural analysis is also necessary to assist the designer in optimizing a number of components whose performance is dependent on their mutual distortions under load (thermal or mechanical). Such systems include the piston/cylinder assembly and valve/valve seat interaction.

3. MAJOR COMPONENTS REQUIRING ANALYSIS

The majority of engine components are subjected to significant loads of some description. However, concentration will be on the major components of the engine, failure of which will either cause total engine failure or serious leakage leading to prolonged engine downtime. These components are

1. Cylinder block:
    This component must maintain structural integrity, provide correct support for the crankshaft and an acceptable environment for the piston.
2. Crankshaft:
    At the heart of the engine is the crankshaft. This component is subjected to high bending loads, torsional loads and centrifugal loading. Adequate fatigue strength must be ensured at the engine concept stage.
3. Connecting rod:
    Located between the piston and the crankshaft, the connecting rod is subjected to high compressive and tensile loadings due to both firing conditions and mechanism inertia.
4. Cylinder head:
    This component is probably the most complex from a geometric viewpoint and is subjected to a range of loadings from thermal to gas pressure.
5. Cylinder head gasket:
    As the cylinder head gasket is located between the head and the block, it is subjected to loadings, both direct from the cylinder gases and those imposed by assembly, and to the effects of other loading in the block and head structure.
6. Piston:
    Probably one of the most critical components in the engine, the piston is subjected to thermal loading, gas pressure and dynamic loads. In addition, it must be of an acceptable shape to allow lubrication between itself and the cylinder bore to be maintained under these loadings and their resultant distortions.

It can be seen from the above that there are strong interactions between the loadings of all these components. The whole engine must be treated as a system!

4. GENERAL MODELLING CONSIDERATIONS

In section 3 above it was pointed out that an engine must be treated as a system. Ideally this means that all components should be analysed simultaneously in order that the interactions between various components are correctly computed.

In practice, however, some modelling simplifications may be tolerated in order that computations become tractable on today's computers. This necessitates the division of the FE analysis task. Below is described the necessary modelling loading and solution phases of a range of typical analyses covering temperature, stress and vibration analysis of various components including pistons and cylinder blocks.

4.1 The Finite Element Models
There are two main classifications of Finite Element used in engine analysis:

 (i) Solid (3D) elements.
(ii) Plate bending elements (2D – but available in 3D space).

For the purposes of stress and temperature analysis the use of solid elements is recommended. In particular the isoparametric 20 node brick element and the 15 node wedge will be considered here (Fig. 1).

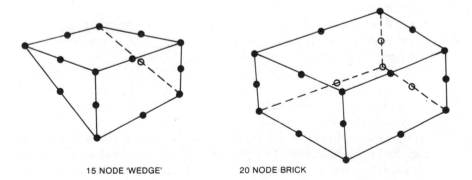

15 NODE 'WEDGE'   20 NODE BRICK

Fig. 1 – 3D solid finite elements.

The first step in the engine modelling process is to identify the major static components. A typical example is shown on Fig. 2 which indicates diagrammatically the cylinder head, gasket cylinder block and crankcase of an engine. This

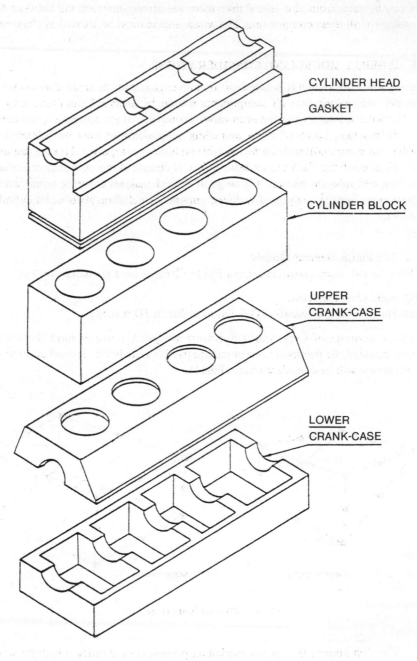

Fig. 2 – Structural components.

diagram, for convenience, subdivides solid structures into more manageable sub-components (not necessarily at physical boundaries). Further subdivision is illustrated on Fig. 3. At the FE modelling stage these subassemblies may be even further divided. Figure 4 gives a typical example of the basic FE meshes used to simulate a slice of an engine. These small assemblies of elements are known as substructures which, when appropriate, may be used to assemble a whole engine model using an appropriate Finite Element substructuring package.

Typically, each substructure will contain about 120 elements which, for stress analysis, will have typically 5000 degrees of freedom. The combined model of the complete block and head structure will be equivalent to a structure of around 250 000 freedoms.

In addition to the 'static' components, an engine analysis will include a piston model (Fig. 5), the connecting rod and a number of crankshaft sub-structures (Figs. 6, 7 and 8). Typical piston meshes will contain between 100 & 300 elements to model one quarter of the piston and gudgeon pin.

The generation of these models is effected by the defining in 3D space of the co-ordinates of all nodes in the structure and the specification of which nodes belong to each particular element. By manual effort the preparation of a cylinder block slice model as depicted in Fig. 4 may take 10—20 man-weeks to prepare (including loading data). In engine structures, conventional 'mesh generators' are of little value in assisting the FE modeller; however, more sophisticated software is now coming to the market-place, which attempts to tackle the difficult problem of idealizing complex structures with reduced manual effort.

Although solid elements are usually chosen for stress and temperature analyses, the preferred element for vibration prediction on engine blocks is the plate element. This family of elements is 2D in geometry and has an assigned thickness property. In addition to their in-plane (membrane) stiffness, out-of-plane motion is controlled by an accurate bending stiffness formulation. Figure 9 shows the commonly used plate bending elements.

The areas of an engine which are usually analysed are the cylinder block structure and the major noise emitting covers of the engine, e.g. the sump, timing gear and valve covers. Examples of typical plate element models are shown in Figs. 10 and 11. The cylinder block model typically contains 850 nodes and 1500 elements; the sump model shown contains 300 elements.

It should be noted that in vibration analysis for noise prediction, the outer surface out-of-plane motion is of prime importance. Should solid elements be used (with the aspect ratio dictated by model size, computer space and run time considerations) the accuracy of this out-of-plane motion would be inadequate — the bending stiffness of high aspect ratio 3D solid elements being very poor.

Care should be taken, when using plate elements to model cylinder block structures, that great attention is paid to accurate definition of plate intersections, where, physically, structures are more of a 'solid' nature.

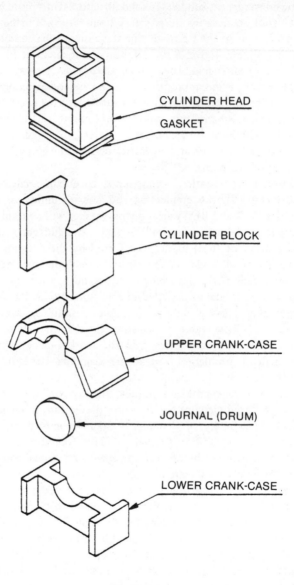

CYLINDER HEAD

GASKET

CYLINDER BLOCK

UPPER CRANK-CASE

JOURNAL (DRUM)

LOWER CRANK-CASE

Fig. 3 — Back-to-back panel.

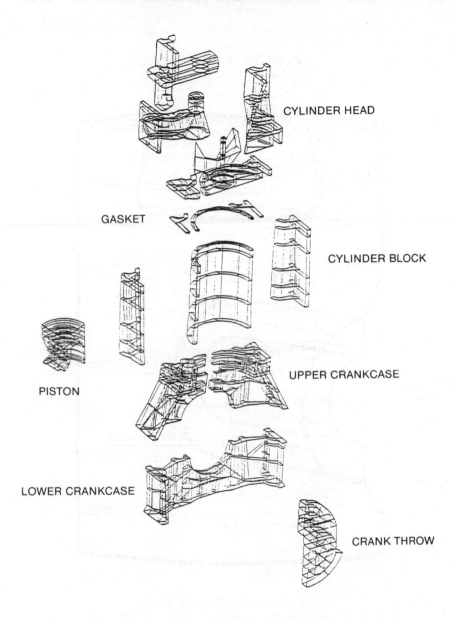

CYLINDER HEAD

GASKET

CYLINDER BLOCK

PISTON

UPPER CRANKCASE

LOWER CRANKCASE

CRANK THROW

Fig. 4 — Substructures used for generating the full engine finite element model.

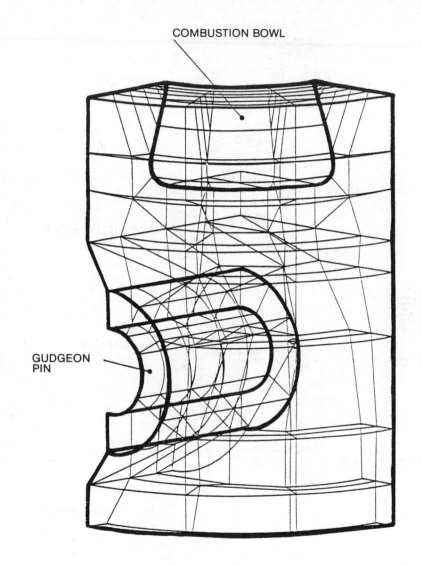

COMBUSTION BOWL

GUDGEON
PIN

Fig. 5 – 3D finite element model.

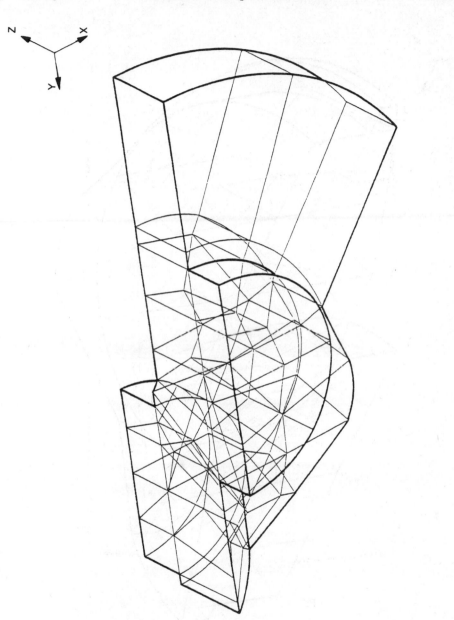

Fig. 6 – Webs 1 and 8.

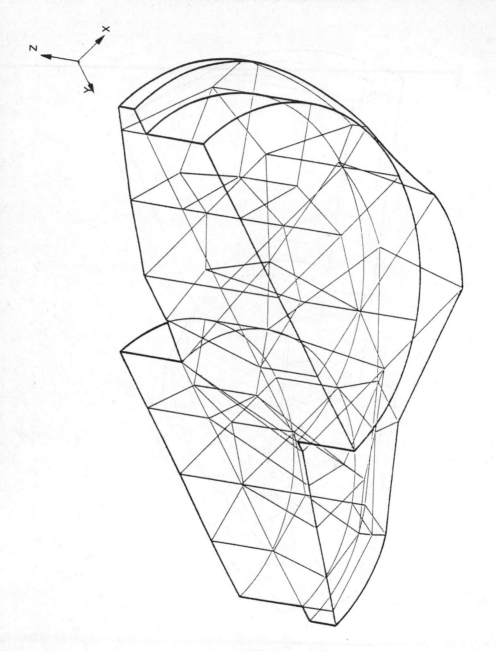

Fig. 7 – Webs 2, 3, 6 and 7.

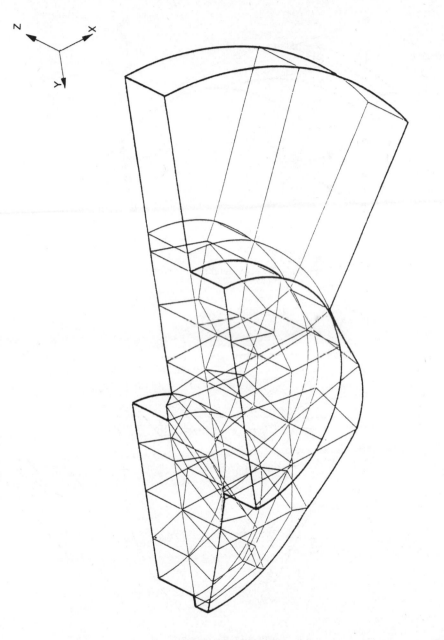

Fig. 8 — Webs 4 and 5.

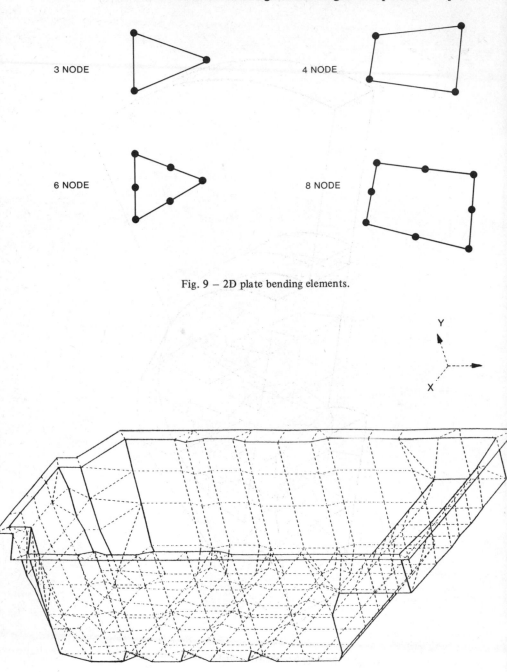

3 NODE

4 NODE

6 NODE

8 NODE

Fig. 9 – 2D plate bending elements.

Fig. 11 – Sump FE mesh.

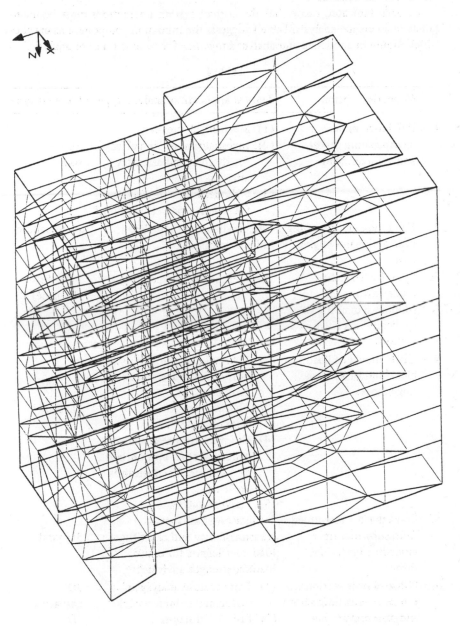

Fig. 10 – Cylinder block FE mesh.

4.2 Engine subsystems

It has been indicated above that the correct system interactions must be maintained in an engine analysis. Table 1 suggests the minimum component assemblies which should be modelled and analysed *together* for various types of analysis.

Table 1

	Model description	Typical application/analysis type	Element type
1.	Full block model incorporating gasket and cylinder head	(1) Mechanical stressing (2) Investigation of 'end effects' for thermal and assembly analysis (3) Cylinder block flexibilities for bearing load calculations*	3D solid
2.	Full block incorporating head and major engine covers	Forced vibration modelling for noise prediction	Plate bending
3.	Cylinder head slice model with valves etc. (double symmetric or cyclically repetitive section)	Head temperature prediction	3D solid (Laplace)
4.	'Slice' model of head, block and gasket (e.g. single bore)	(1) Thermal distortion of cylinder bore, head, valve/valve seat geometry, etc. (2) Thermal stresses (3) Assembly stresses (4) Gas pressure stresses on head/gasket (if appropriate anti-symmetric constraints are used) (5) Gasket analysis	3D solid
5.	Crankshaft — assembled from substructure modelling individual throws	Evaluation of stiffness, transmissibility and centrifugal load coefficients for crank loading/strength analysis*	3D solid
6.	Piston ¼ or ½ section + ¼ or ½ of cylinder bore and upper and lower block decks	(1) Temperature analysis of piston and cylinder block (2) Piston distortion (3) Piston stresses — thermal gas loading etc.	3D Laplace 3D stress

*See Chapter 6 on bearing load calculation.

5. LOADINGS

The accurate description of engine loadings is a prerequisite to obtaining correct predictions of stress, temperature or vibration levels. This section will cover the two main categories of loading experienced by engine components, i.e. thermal and mechanical. Section 5.2 will cover the effects of gas pressure, bearing loads, etc. and assembly loads.

5.1 Modelling the Thermal Loads — General

The engine analyst must first recognize the relevant heat transfer mechanisms present in an engine:

(1) convection,
(2) conduction,
(3) boiling,
(4) 'contact resistance',
(5) radiation.

Convection is present in all areas of the engine where there is a flow of gas such as in the cylinder/combustion chamber and in the inlet and exhaust ports. The cooling of the engine by either water or air systems is dominated by convection as is the cooling of the piston to the in-crankcase gases and to the cylinder walls. The usual convection correlations may be found in the literature, e.g. [1,2].

Conduction is the mechanism for heat flow in the engine structure itself, i.e. the head, block and piston.

Boiling may be found in the high temperature areas of an engine water jacket, particularly adjacent to the top of the cylinder bore and the valve bridge.

For typical correlations see [1, 2].

'Contact resistance' — heat transfer from a number of components in the engine are affected to greater or lesser amounts by contact resistance. The major areas of interest, where critical component temperatures are affected, are (i) the cylinder liner/parent bore interface of dry liner engines and (ii) valve-to-valve seat and seat insert to head [3].

Radiation may be the major mechanism of heat loss from exhaust manifolds and is a minor mechanism of heat transfer from exhaust gases to the exhaust system. In low swirl type combustion systems, significant heat transfer from in-cylinder gases to the combustion chamber walls may complement convection.

The Finite Element engine analyst must have an appropriate knowledge of the above heat transfer mechanisms to choose representative values of the heat transfer coefficients which surround the engine component.

The analyst may choose one of the well known correlations for in-cylinder heat transfer such as

(1) Eichelberg.
(2) Annand (*Inst. Mech. Eng.,* **177**, No. 36 (1963)).
(3) Woschni (MTZ, 1970).

The equations due to the above will normally be incorporated into a 'cycle' simulation program.

As an example, the Woschni equation is described below:

$$h = 110D^{-0.2} P^{0.8} T^{-0.53} \left[C_1 C_m + C_2 \frac{V_H}{P_1} \frac{T_1}{V_1} (P - P_0) \right]^{0.8}$$

where
   h is the heat transfer coefficient
   D is the cylinder bore diameter
   T is the in-cylinder gas temperature
   C_1 is (a) during gas exchange $= 6.18 + 0.417 \, C_u/C_m$
        (b) during compression and expansion $= 2.28 + (0.308 \times C_u/C_m)$
   C_u/C_m is swirl ratio
   C_2 is a combustion process dependent constant
   V_H is stroke displaced volume
   $P_1 V_1 T_1$ are conditions at start of compression which define the trapped mass.
   $P - P_0$ is the pressure difference between the 'fired' and 'motored' engine.

Convention in other areas of the engine will frequently require the use of a standard Nusselt/Reynolds type correlation, e.g. $Nu = A Re^{0.8} Pr^{0.33}$ where A will vary from $0.023 - 0.05$, depending on how 'developed' the flow is considered to be.

Figure 12 gives a table of some typical values of HTC for the various mechanisms found on engine components. Additional information may be found in the sections of the chapter specifically covering component temperature calculation.

5.2 Mechanical Loads
The major loadings which are considered in engine analysis are identified below:

(1) bearing loads,
(2) piston side thrust (effect on both block and piston),
(3) cylinder head/piston gas load,
(4) inertia loads on piston,
(5) connecting rod inertia loads,
(6) cylinder head assembly loads,
(7) main bearing cap assembly loads.

For the purposes of cylinder block stressing, reference should also be made to Chapter 6 on crankshaft and bearing performance analysis. This chapter, which also covers crankshaft stress analysis, identifies the relevant dynamically induced loadings on the engine (loadings 1, 2 and 3, above).

The static loadings (6 and 7) which are applied to cylinder blocks are defined by the necessary tightening torque to achieve satisfactory joint performance, e.g.

Component cooling

Component	Mechanism	Typical values
Piston	To liner:	
	Convection/conduction	0.75–1.25 mW/mm² °C
	To U/C:	
	Oil mist	0.75–1 mW/mm² °C
	Jet cooling	1.0–3.0 mW/mm² °C
	Cocktail shaking	1.0–3.0 mW/mm² °C
		(Bush and London)
Head and block	Convection	$Nu = A Re^{0.8} Pr^{0.33}$
		$A = 0.023$–0.05
		Min value 5.0 mW/mm² °C
	+ Boiling	Max value 10.0 mW/mm² °C
Liner/block	Contact	
Valve/seat	Resistance	Pressure, hardness and roughness dependent 3.0–6.0 mW/mm² °C [3]

Fig. 12 – Component cooling.

butt face stress condition, gasket sealing loads. The macro effect of these loadings may be applied by the use of equal and opposite forces applied to the FE model in the bolt head and thread regions. However, close examination of the loaded area itself will require the detail of the fixing to be modelled.

Inertia loadings are calculated from component accelerations which are evaluated from the piston/rod crank mechanism geometry and are simulated by 'body force' modelling in suitable FE packages.

The importance of some of the above variables will be observed in section 6, which deals with specific analysis examples.

6. TYPICAL ANALYSES

This section is devoted to describing the detail of models and loading procedures for a number of the sysbsystems identified in section 4.2. It should be noted that some analyses require the results of other model subsystem analyses as part of their load input, e.g. cylinder head temperature distributions will be used in stress analysis of the head/block/gasket assembly.

Each of the following sections will in general identify the type of analysis, specific model requirements, loadings and constraints together with some typical results. The subsystems discussed are not intended to be a full list of the analyses required to theoretically validate an engine structure. They merely indicate the approach and nature of the necessary analyses.

The analyses discussed are centred around the use of a Finite Element analysis program with substructuring facilities. It is pointed out that although such facilities are *not essential,* use of a program without a substructuring package will lead to wasted modelling and CPU time and, in some instances, demand some very, very long computer runs.

6.1 Piston and cylinder temperature analysis
6.1.1 The FE model
In order to compute correctly the temperatures in a piston, the environment of the component must also be modelled. In this case the surrounding cylinder block structure must be included in the temperature analysis model as the cylinder bore temperature is not independent of piston temperatures. In Fig. 13, a typical 3D FE model of a piston is illustrated. The model of the cylinder liner, parent bore, cylinder block top and bottom decks and outer side wall is illustrated in Fig. 14. The 'one-quarter' model structures are the bare minimum that need be included in the temperature analysis. In this case the structure must be assumed to be doubly symmetric and also that there is a negligible heat loss from the block top and bottom decks.

Although acceptable 2D models could be generated for the piston, the necessity for a knowledge of component temperature distribution in 3D for stressing purposes and assessing the 3D distribution of temperature typical on cylinder barrels (particularly in the case of a siamesed bore engine) is sufficient justification for a 3D temperature calculation.

6.1.2 Thermal loading
The evaluation of accurate and consistent temperature distributions requires a knowledge of the thermal loading applied to all surfaces of the piston and cylinder Finite Element models. The level of in-cylinder heat transfer coefficient is derived by the use of a cycle simulation program incorporating Woschni's correlation (see section 5.1). The equations expressing the variation of time—mean heat transfer coefficient and gas temperature, as seen by the piston crown and cylinder liner, are given in equations (1)–(4), and are illustrated diagrammatically in Fig. 15. This fig. also shows the spatial variation of crown heat transfer coefficient typical of toroidal bowl pistons.

$$\bar{h}_g = \frac{\int_0^{720} h_g \, d\theta}{720} \tag{1}$$

$$\bar{T}_g = \frac{\displaystyle\int_0^{720} h_g T_g \, d\theta}{\displaystyle\int_0^{720} \bar{h}_g \, d\theta} \tag{2}$$

where h_g and T_g are the heat transfer coefficient and gas temperature at crank angle θ, respectively.

The value of $\bar{h}_g$ 'seen' by the piston crown will in general be specific to a particular combustion bowl shape and consequently appropriate scale factors must be applied to Woschni heat transfer coefficient predictions to suit specific combustion systems. Similarly the distribution of heat transfer coefficients (spatially) on the crown is generally determined by experiment for a range of typical bowl geometries.

The distributions of effective time—mean gas temperature and heat transfer coefficient between gas and cylinder are evaluated from equations (3) and (4), below.

$$\bar{T}_{g(\theta_1)} = \frac{\displaystyle\int_{\theta_1}^{360-\theta_1} h_g T_g \, d\theta + \int_{\theta_1 + 360}^{720-\theta_1} h_g T_g \, d\theta}{720 \, \bar{h}_{g(\theta_1)}} \tag{3}$$

$$\bar{h}_{g(\theta_1)} = \frac{\displaystyle\int_{\theta_1}^{360-\theta_1} h_g \, d\theta + \int_{\theta_1}^{720-\theta_1}{}_{360} h_g \, d\theta}{720} \tag{4}$$

where $\bar{T}_{g(\theta_1)}$ and $\bar{h}_{g(\theta_1)}$ represent the effective mean value of gas temperature and heat transfer coefficient as experienced by a point on the cylinder which becomes exposed to the in-cylinder gases at crank angle θ_1.

The heat transfer coefficients quoted for the piston under-crown, water jacket and the cylinder/parent bore interface are typical values derived from published correlations and research conducted at the author's company.

To complete the boundary condition data, the dynamic thermal interaction of the piston, the rings and the cylinder liner is approached using a quasi-static approximation. Figure 16 illustrates diagrammatically a Finite Element mesh of the piston and liner interface. A connectivity matrix can be constructed which identifies the liner elements that are traversed by each element on the piston skirt between TDC and BDC. During the overlap of particular piston and liner elements, steady state heat transfer is assumed. The dynamics of the system can be included by weighting the individual heat transfer coefficients according to the appropriate overlap times of piston and liner elements. These 'dwell' times can be evaluated by reference to a piston/crank angle displacement diagram.

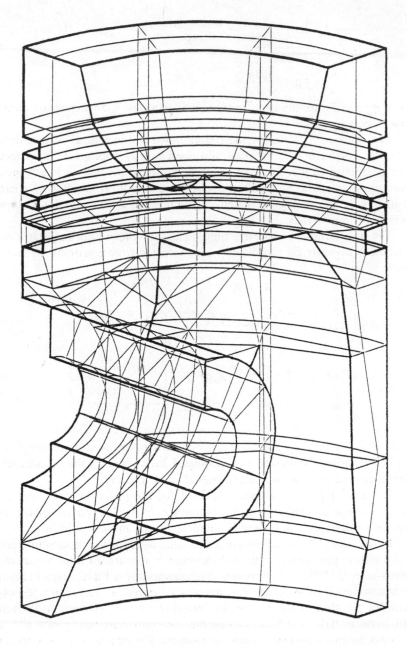

Fig. 13 – FE mesh of piston.

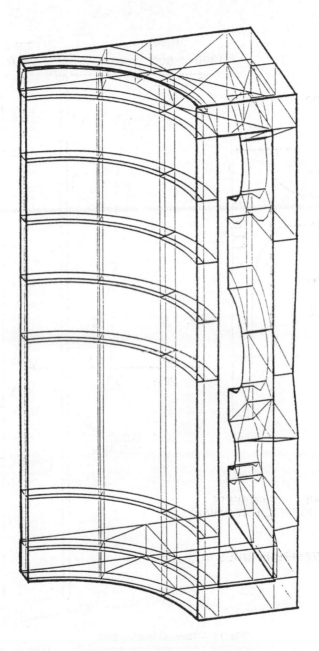

Fig. 14 — FE mesh of cylinder (temperature analysis).

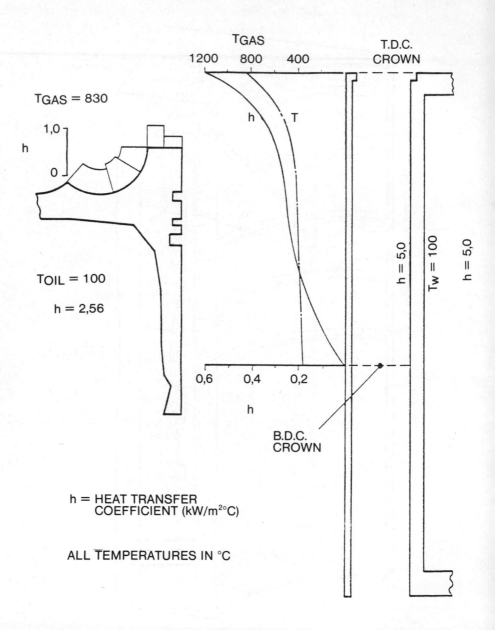

Fig. 15 – Thermal loading data.

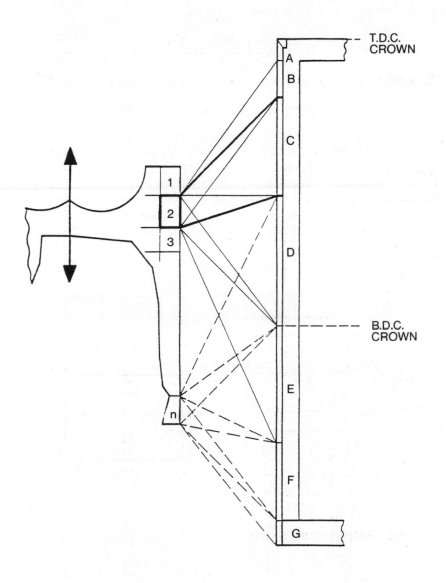

Fig. 16 — Piston/cylinder thermal connectivity matrix.

Thus the time average heat transfer coefficient between piston element 2 (Fig. 16) and liner element C is expressed by

$$\bar{h}_{(2-C)} = h_{(2-C)} \frac{\Delta\theta}{180} (2-C)$$

where $\Delta\theta_{(2-C)}$ is the passing time of element 2 over element C. Clearly,

$$\Delta\theta_{(2-A)} + \Delta\theta_{(2-B)} + \ldots + \Delta\theta_{(2-D)} = 180°$$

The full matrix of heat transfer coefficients between all surface elements on the piston (and rings) and the cylinder liner is represented by Boundary Layer Elements in the PAFEC Finite Element System [4].

The distribution of heat transfer on the skirt of a piston is typified by the distribution shown in Fig. 17. Also shown in this figure are characteristic values thar are to be found around the ring. These coefficients control the heat flow into the ring from the piston ring groove and also out of the ring to the liner.

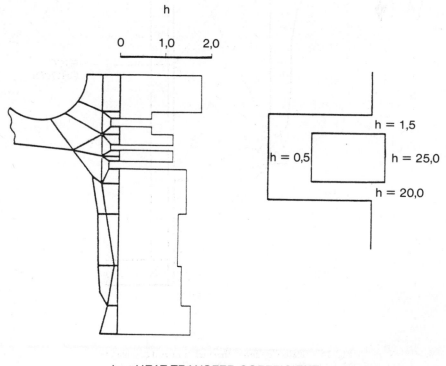

h = HEAT TRANSFER COEFFICIENT
(kW/m$^{2°}$C)

Fig. 17 – Distribution of piston skirt heat transfer coefficient.

6.1.3 Typical Temperature Predictions

The results of a temperature prediction using the approach identified above are summarized in Fig. 18 for an engine operating at rated power.

The accuracy of this prediction is confirmed by thermocouple measurements, the results of which are also given in Fig. 18.

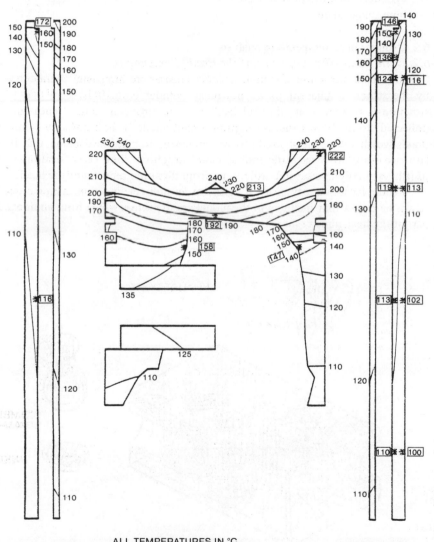

ALL TEMPERATURES IN °C
* □ MEASURED TEMPERATURES

Fig. 18 – Computed piston and cylinder temperature distribution.

It will be usual to store on file the temperatures computed in the piston, gudgeon pin, cylinder and cylinder block as they are generally used for the following

— piston thermal stressing
— piston thermal distortion
— cylinder block thermal stressing
— bore shape analysis

6.2 Cylinder head temperature analysis

6.2.1 The Finite Element model of the Head (Temperature)

Cylinder head structures on multi-cylinder engines are in general characterized by the structure adjacent to an individual cylinder bore. It is usual that this structure is present along the full length of the engine in either 'repeated' or 'reflected' form. Hence the assumption usually made is that only one cylinder bore's worth of the head needs to be modelled. It may be debated that it is also necessary to investigate end cylinder structures, but these sections will usually vary only in a second order sense from the other sections of the head.

Figure 19 shows a typical FE mesh for an indirect injection cylinder head. It will be noted that included in the model are the *valves* which form an integral part of the heat transfer model.

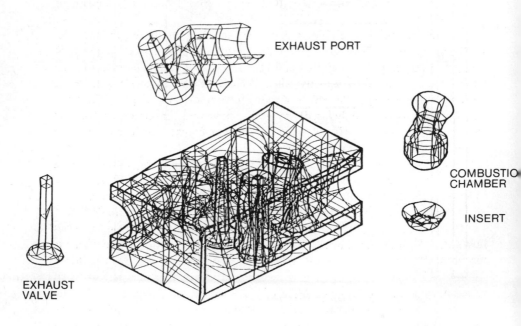

Fig. 19 – Finite element mesh: cylinder head.

The porting configuration on the cylinder head will determine the type of 'constraint' required on the cut faces to simulate correctly the heat flow along the engine. In the case of a 'repeated' geometric structure this is effected by 'constraining' the temperature distribution of both cut faces of the head to be identical at corresponding points. Symmetric structures may be considered to be insulated on the cut faces. Typical configurations for a four-cylinder engine are

EI	IE	EI	IE	— symmetric structure
IE	IE	IE	IE	— repeated structure

6.2.2 Thermal loading

The thermal loading of the engine is, as in the case of the piston, initiated by conducting the appropriate cycle simulation program runs to evaluate in-cylinder heat transfer coefficient and gas temperature averages (see sections 6.1.2 and 5.1) for the engine operating condition under consideration. Knowledge of the perturbations of the mean heat transfer coefficient — if available from typical engine tests — allows the thermal loading on the flame face of the cylinder head and valve faces to be fully described. Figure 20 illustrates typical heat transfer coefficient information. The area of high heat transfer coefficient adjacent to the combustion pre-chamber results from the high gas velocities in this region of the flame face.

The heat transfer coefficients in the water jacket of the engine must be evaluated by the use of convection coefficient correlations or boiling heat transfer relationships — see Fig. 12 and [1] or [2]. These generally require some knowledge of water flow velocity which, in the case of a paper engine study, must be evaluated with the aid of a water flow prediction program (the pipe/reservoir type is commonly used). Typically, however, convection coefficients will be in the range $5-10$ mW/mm^2 °C whereas boiling coefficients may be almost an order of magnitude higher!

The contact resistance between the cylinder head and the valves may be obtained from a suitable correlation — see [3] and Fig. 12, and must be 'scaled' to account for the time when there is no contact during valve-open periods.

Typically, the coefficients in this region will be about $3-6$ mW/mm^2 °C.

Cycle simulation program predictions are used again to determine port gas flow rates. Appropriate convection correlations are used to compute coefficients for both the port flow and the valve/valve seat areas of the cylinder head. Appropriate time scaling is again necessary to account for time periods when valves are closed and no flow takes place.

Heat transfer from the head to the gasket is assumed to be zero as (a) the top of the block is generally at a similar temperature to the bottom deck of the head and (b) gaskets are generally constructed of insulating materials.

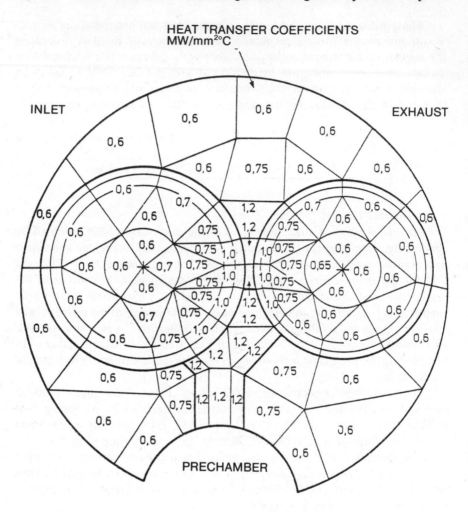

Fig. 20 – Engine cylinder head. Distribution of flame face heat transfer coefficients.

Source and sink temperatures other than the mean gas temperatures discussed above are: the water coolant temperature, inlet air temperature in the inlet port, and exhaust temperature in the exhaust port. The back faces of the valves and the seats are also subjected to these temperatures.

6.2.3 Typical temperature predictions in the cylinder head

The temperature results of a typical analysis are illustrated in Figs. 21, 22 and 23. These include both temperature isotherm contours on both the flame face and the section through the cylinder head.

As for the piston analysis of section 6.1, temperature information will usually be stored for later stress and displacement analysis, e.g. flame face thermal stresses, valve/valve seat mismatch assessments and gasket analysis.

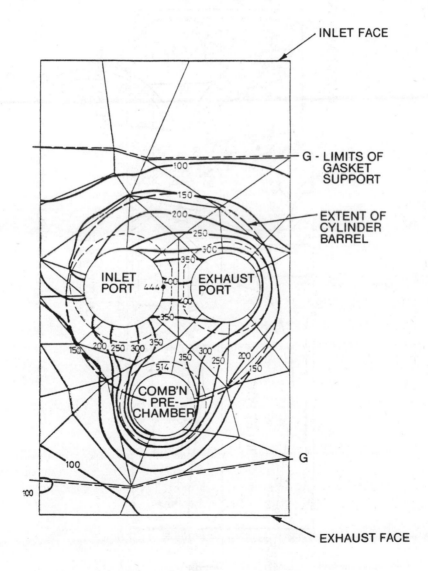

Fig. 21 – Engine cylinder head. Flame face temperature distribution.

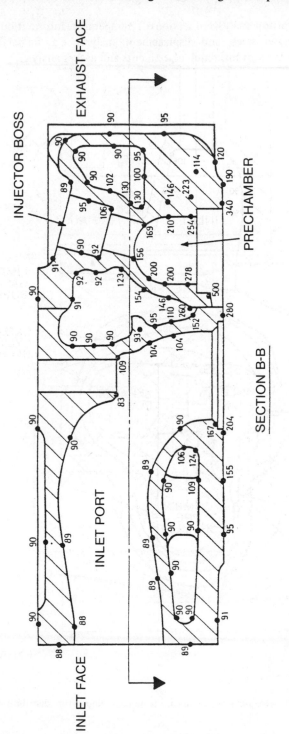

Fig. 22 – Engine cylinder head. Predicted temperatures °C.

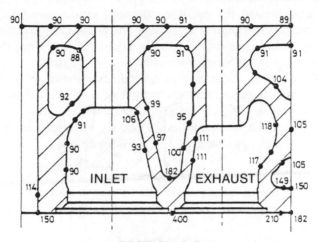

SECTION D-D

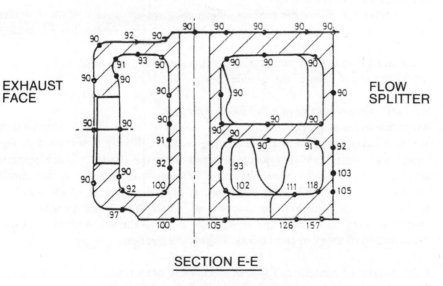

SECTION E-E

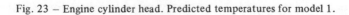

Fig. 23 – Engine cylinder head. Predicted temperatures for model 1.

6.3 Cylinder block/gasket/head mechanical stress analysis
This section covers the most costly FE modelling aspect of engine analysis and will in general demand the use of a multi-level substructuring facility in the Finite Element program Mechanical loading will primarily be considered here.

Integrity of an engine structure can be computed with a knowledge of the various cyclic and steady state stresses imposed on the structure by the service loading of the engine. These will in general comprise assembly, thermal and 'mechanical'. The mechanical loadings themselves result – as indicated in section 5.2 from gas pressure and inertia effects. Clearly, as mechanical load 'symmetry' or repeatability cannot be assumed in the engine structure, for mechanical stress evaluation a full engine structure must be considered.

6.3.1 Finite Element model – full engine structure
The major structures of an engine are shown diagrammatically in Fig. 2. They will in general comprise the cylinder block, the crankcase, the gasket and the cylinder head.

The first step in generating the engine model is to subdivide these structures – if possible into a representative, repetitive section of the engine. Usually this will be a section containing two half-bores and one main bearing cap. Figure 3 illustrates a typical example.

At this stage, Finite Element meshes of each sub-component of the structure are generated. This may require further subdivision of the engine. The FE meshes of each of these substructures, as they are known, are shown in Fig. 4.

Extra FE models of non-repeated structures must also be generated, e.g. to represent the end faces of the cylinder block.

6.3.2 Mechanical loading of a full block structure
An engine cylinder block is subjected to a series of 'mechanical' loadings which are time dependent. The major loads are discussed briefly in section 5.2. The largest and most important loadings result from the cyclic dylinder pressure applied to the cylinder head, the main bearing loads and piston side thrust loadings. The topic of bearing loads etc. is fully discussed in Chapter 6. It is sufficient to say here that the time history of all these loads, including their relative phasing throughout the engine, should be evaluated in order to ensure computation of the correct mechanical stress distribution.

6.3.3 Solution technique and typical mechanical stress results.
The necessary solution strategy is described below:
(i) Section 6.3.1 indicated the geometry of each unique Finite Element sub-structure from which the cylinder block model is to be constructed. The first task in the generation of the full finite element models is to identify the nodes on each substructure which are to connect to other substructures. Appropriate freedoms at these nodes together with those at any nodes

which are subjected to mechanical loading, including mounting points, must
be held as 'master' degrees of freedom.

(ii) Having obtained individual stiffness matrices, these may be 'duplicated',
reflected and/or translated in space to allow the assembly of all substructure
components of the cylinder block assembly. Using the frontal solution
technique, care must be taken in the initial mesh designs to ensure that an
acceptable front size is produced during this assembly stage of all substructures to form the full block. Again, during this stage of the analysis,
master freedoms must still be retained at all loaded nodes and mounting
points of the engine.

(iii) When the substructure assembly has been completed and all internal freedoms
reduced, the remaining engine stiffness matrix will contain only those
freedoms where loads are applied and where constraints need to be placed
to model mounting points (see Fig. 24).

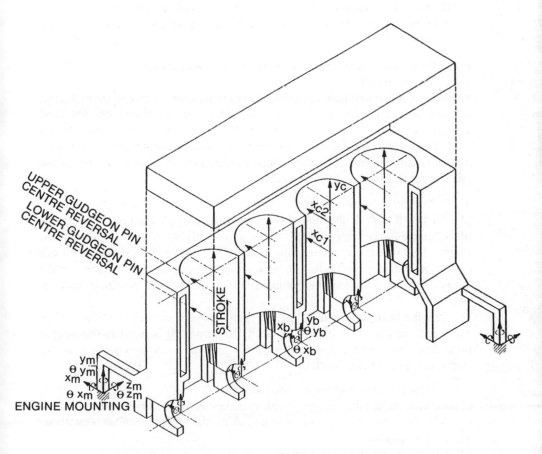

Fig. 24 – Illustration of engine force and displacement directions.

(iv) Having selected *all loadings* corresponding to a particular crank angle, and constraining the mounting point freedoms, the FE program can solve the remaining stiffness equations to yield the *displacements* at the load application master degrees of freedom.

The back substitution process allows displacements at each substructure boundary to be computed. Repeating the back substitution process allows the displacements and stresses at the nodes in every element to be evaluated.

(v) In order to obtain the full block stressing picture, step (iv), above, must be repeated to obtain full stress maps for all loadings at a series of crank angles from $0 - 720°$.

Examples of such stress plots for sidewall core holes are shown in Figs. 25 and 26. As fatigue life assessment is probably the reason for conducting stress analyses of this type, an alternative method of presentation may be adopted which displays the stress at a given area of interest as a function of crank angle. Figure 27 shows the cyclic nature of the stress at a core hole. Experimental results from a running engine are also shown for comparison.

6.4 Cylinder Block/head/gasket assembly and thermal stresses
6.4.1 The FE model.
Section 6.3 illustrated how cyclic mechanical stresses in an engine are evaluated. In order to conduct fatigue life assessments etc. the effect of assembly loads and thermal loads must be considered. Clearly, full engine models will cater for the above loading, but usually only a representative *slice* of the engine need generally be considered. (Figures 28 and 4). Such models will, of course, need to contain cylinder block gasket and head sections.

6.4.2 Thermal and assembly loads
In sections 6.1 and 6.2, the method of computation of cylinder head and cylinder block temperatures was described. These data — stored on file — are used as the thermal loading data for the chosen rating of the engine. In addition, the loads applied by cylinder head bolts etc. may be applied as a separate load case to the model in order to allow the computation of thermal and assembly loads in isolation.

The choice of a slice model of the engine necessitates the use of a suitable constraint condition on the cut faces of the engine. It is usual to ensure that both these faces remain plane and parllel — this allows axial expansions of the engine but limits engine bending.

6.4.3 Typical results from engine slice models.
Clearly, stresses in both the block and cylinder head due to the applied loadings are available; however, only two examples of the use of the model are described:

(i) Gasket analysis.
(ii) Cylinder bore distortion.

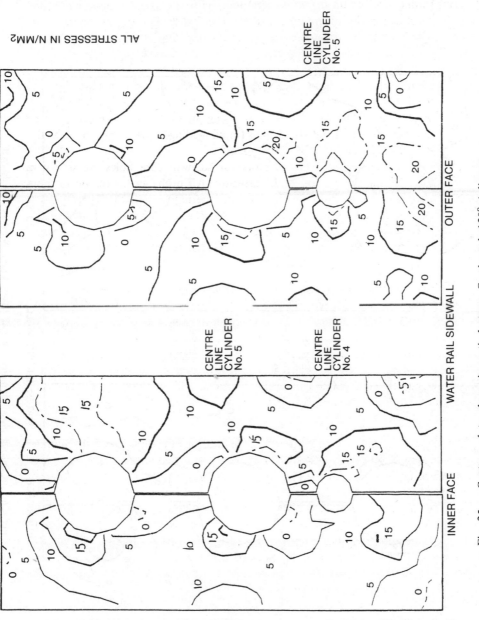

ALL STRESSES IN N/MM2

CENTRE LINE CYLINDER No. 5

OUTER FACE

WATER RAIL SIDEWALL

CENTRE LINE CYLINDER No. 5

CENTRE LINE CYLINDER No. 4

INNER FACE

CENTRE LINE CYLINDER No. 4

Fig. 25 — Contour plots: dynamic vertical stress. Crank angle-120° (adjacent cylinder no. 5 firing).

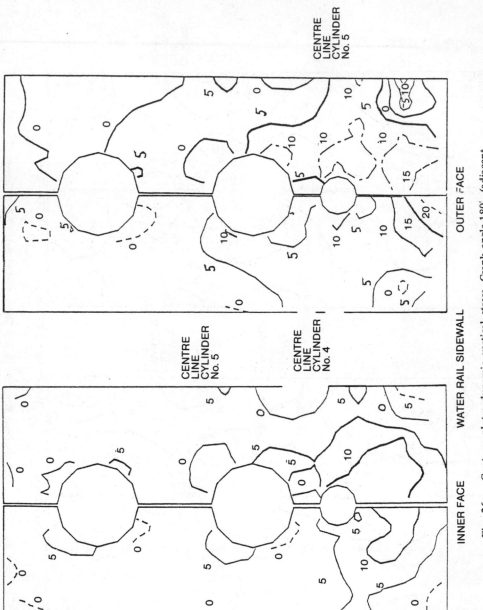

Fig. 26 — Contour plots: dynamic vertical stress. Crank angle-180˚ (adjacent cylinder no. 5, 60° after firing).

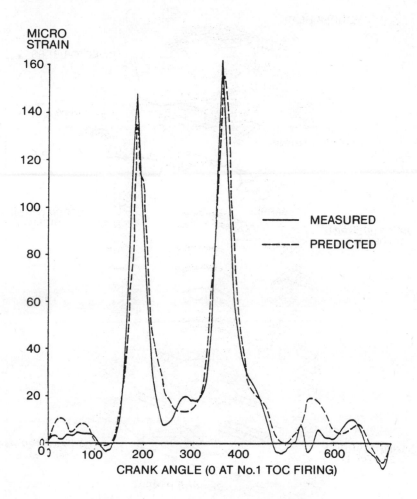

MICRO
STRAIN

CRANK ANGLE (0 AT No.1 TOC FIRING)

MEASURED

PREDICTED

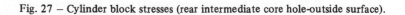

Fig. 27 – Cylinder block stresses (rear intermediate core hole-outside surface).

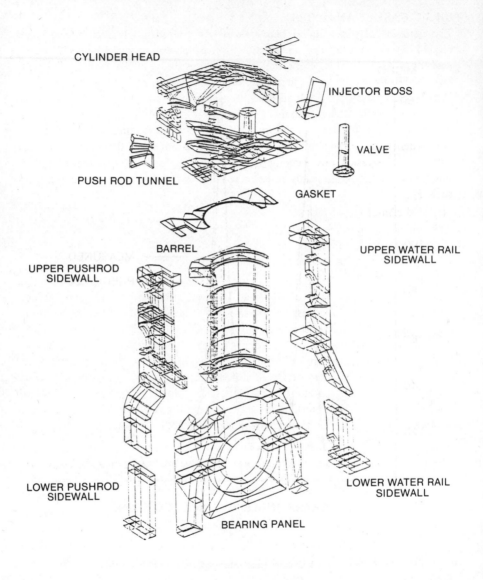

CYLINDER HEAD

INJECTOR BOSS

VALVE

PUSH ROD TUNNEL

GASKET

BARREL

UPPER WATER RAIL
SIDEWALL

UPPER PUSHROD
SIDEWALL

LOWER PUSHROD
SIDEWALL

LOWER WATER RAIL
SIDEWALL

BEARING PANEL

FINITE ELEMENT MODEL
SUBSTRUCTURES

Fig. 28 — FE substructure meshes (for displacement analysis).

6.4.3.1 GASKET ANALYSIS

The presence of the cylinder head gasket in the head/block assembly leads to modelling complications if an 'elastic' Finite Element program is used for the analysis. This is because both the bead and the body of the gasket behave in both a non-linear and a plastic manner under the influence of compressive loads. Figure 29 shows a typical load/deflection curve. By a process of iteration it is possible to compute the permanent distortion of a gasket by applying the model assembly to the full range of thermal, assembly and gas loadings* to which these components are subjected. The 'E' values consistent with the permanent distortion of the gasket now become the starting data for the assessment of gasket integrity. The model assembly must be stressed under a range of load conditions such as

 (i) cold clamping,
 (ii) cold clamping + gas pressure,
(iii) clamping + thermal,
(iv) clamping + thermal + gas pressure.

Figures 30, 31, 32, 33 indicate typical computed load distributions under the above loadings. Inspection of the bead loadings indicates the change in gasket bead loading due to gas pressure and thermal effects. The strain in the gasket is also indicated on the crankshaft and thrust axes of the bore.

The above information will generally indicate where sealing problems due to leakage are likely to occur and also the loadings which contribute to fatigue. In addition to assisting in the design of this important joint, this information also assists the engineer to develop the best test bed cycle to validate gasket integrity, e.g. is thermal cycling important for accelerating fatigue of the bead or is a gasket vulnerable under cold but high gas load conditions?

6.4.3.2 CYLINDER BORE DISTORTION

The major contributors to cylinder bore distortion are (a) thermal loading which generates differential expansion in the cylinder block/cylinder head assembly and (b) the clamping forces provided by the cylinder head bolts. Computation of the distortions resulting from these loadings is straightforward once component temperature distributions are available and the appropriate gasket stiffnesses have been identified (see section 6.4.3.1). Figure 34 depicts computed distortions on the thrust and gudgeon pin axes of the cylinder bore. Thermal gas pressure (see footnote in section 6.4.3.1) and assembly results are plotted separately. Measured assembly distortions are also shown for comparison. It should be noted that the thermal distortions are dominant in generating bore ovality in the diesel engine.

* Ideally, gas loading information should be extracted from the full engine model but by appropriate use of superposition of loadings and cut plane constraint conditions (symmetric and anti-symmetric), the effect of a single cylinder firing may be approximately simulated. In this case, rigid body motion of the slice model in the cylinder axis direction must be constrained at the main bearing housings.

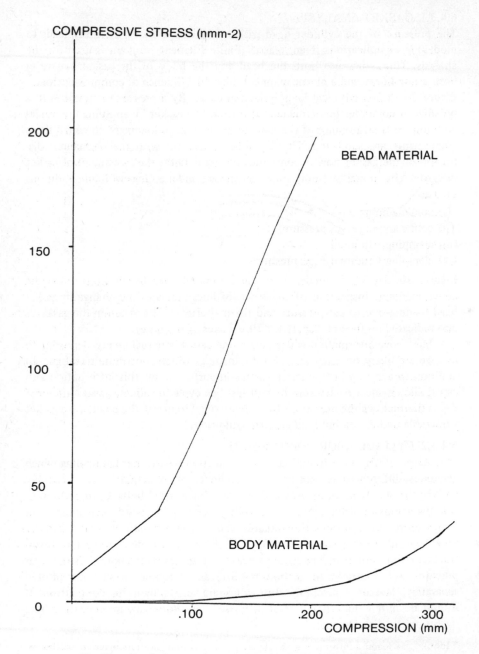

Fig. 29 – Graph showing stress/strain relationship (gasket material).

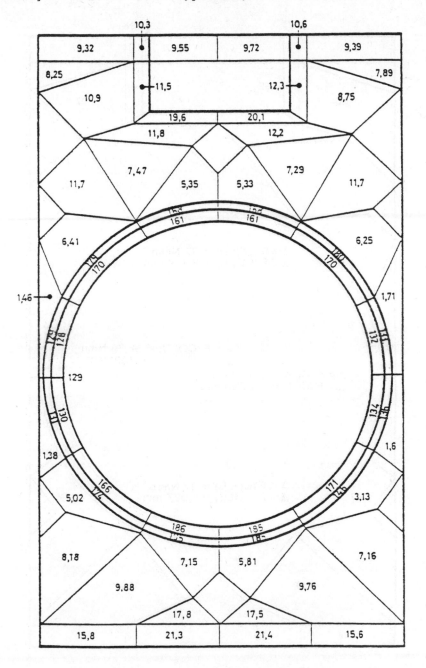

Fig. 30 – Cylinder head gasket: cold clamping.

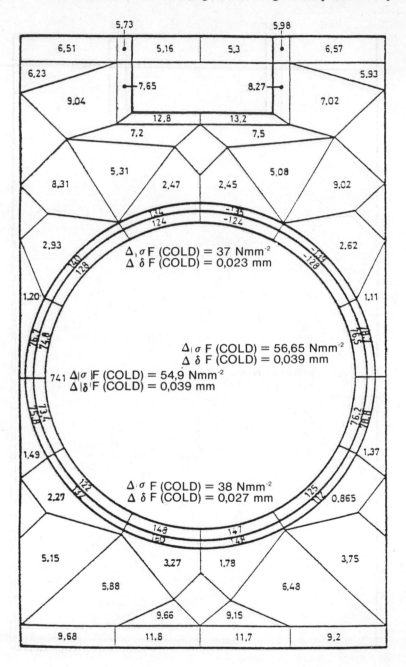

Fig. 31 — Cylinder head gasket: cold clamping and firing pressure.

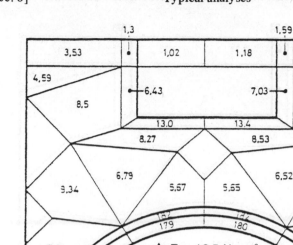
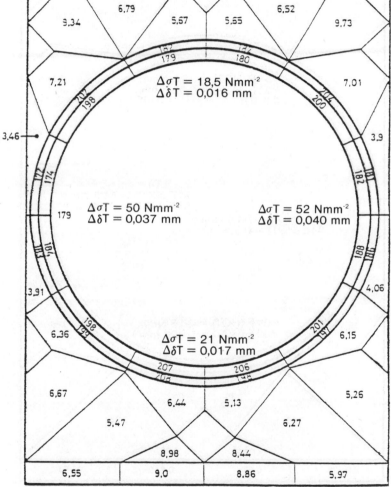

Fig. 32 – Cylinder head gasket: hot clamping.

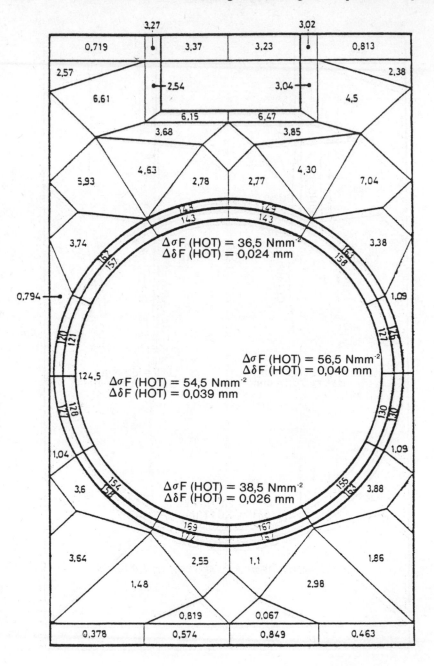

Fig. 33 – Cylinder head gasket: hot clamping and firing pressure.

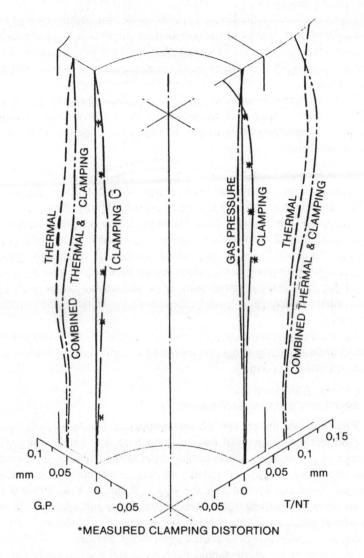

Fig. 34 – Distortion of cylinder bore.

6.5 Piston stress and shape analysis

6.5.1 The FE model

The basic geometry of the Finite Element model for piston stress analysis must be considered with that used for temperature prediction — see section 6.1. Care should be taken, however, to ensure that material properties etc are defined correctly to describe any cast-in inserts, e.g. ring groove armour and expansion control devices.

It will normally be necessary to have available a gudgeon pin model, to be analysed in combination with the piston for certain load condition simulations. Appropriate constraint information is described, together with the loading information, below.

6.5.2 Loading conditions

In this section only thermal and gas pressure loading will be considered. These are the minimum necessary to investigate piston integrity. Thermal loading data, where required, are available from the temperature output of section 6.1. Gas pressure loading will normally be applied to the piston crown and down to the top ring. 'Bimetallic' effects should be applied by artificially raising insert temperatures by an amount equivalent to that necessary to provide equal expansion to that of the base piston material at the known isothermal 'stress free' temperature — taking into account differences in respective component expansion coefficients.

The constraint conditions on a piston in service vary throughout the engine cycle. In addition to any symmetric constraints applied to cut faces of the piston model, it is usual to consider

(1) the 'free' piston and
(2) the gudgeon pin restrained piston.

'Free' analyses are relevant to non-mechanical load cases only, when piston axis rigid body motion need be constrained at any *single* point. Under the influence of gas loadings, when the inclusion of the gudgeon pin is essential, the piston should be constrained radially to follow the pin over a contact arc of about $60°$ from the vertical over the length of the pin boss. (This is typical of the contact angle resulting from piston/pin distortion and a differential expansion induced hot clearance.) Actual constraint extent may be defined after a run by checking for 'tensile' stresses over the constrained surface.

The gudgeon pin itself should be constrained appropriately for symmetry conditions if applicable and to restrain vertical motion where gas loads are reacted on the connecting rod.

6.5.3 Conditions to be analysed

The stress runs normally conducted to assess component integrity are defined below:

(i) 'Free' thermal stress at room temperature to predict cold residual stresses.

(ii) 'Free' thermal at rated power temperatures.
(iii) Pin constrained thermal + gas pressure.

Load cases (iii)–(ii) provide the cyclic mechanical stress range for the hot piston, and load cases (ii)–(i) provide the cyclic thermal stress range.

Some assessment of the variation in mechanical and thermal stresses over the engine duty cycle can be obtained by conducting 'free' thermal runs at 'idle' temperature and pin constrained runs with F/L gas pressure and idle or cold temperature distributions.

The results of the analysis will normally be of interest in the following critical areas of the piston:

(a) bowl lip,
(b) gudgeon pin boss,
(c) ring above armour/parent aluminium interface,
(d) expansion control insert and adjacent areas of skirt.

6.5.4 Piston Shape Analysis

On completion of a Finite Element stress analysis, the piston outer shape under thermal and gas pressure loadings may be investigated. In general it will be found that thermal displacements are an order of magnitude higher than those due to mechanical effects. An example of thermal distortion is shown in Fig. 35.

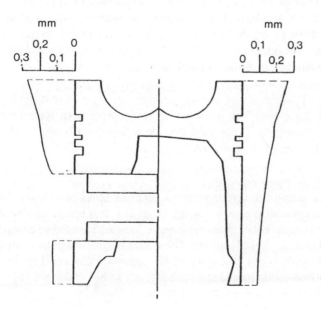

Fig. 35 – Thermal expansion of piston.

The engine analyst should inspect both piston distorted shape and predicted bore distortion prior to engine development in order to suggest a suitable pistion profile for initial testing.

The results of thermal distortion analysis, integrated with piston attitude program predictions, are useful in the assessment of ring groove inclinations during the engine thermodynamic cycle and may be used as data for ring oil film hydrodynamic/oil control analysis.

6.6 Engine Vibration Analysis

In order to assist the engine designer to create a product with low noise generation, an understanding of the forced vibration characteristics of the engine is required. The major noise sources on an engine are usually the sump, timing gear cover, valve cover and cylinder block, the last-mentioned usually being the driving instrument for the covers themselves. An approach to noise prediction based on Finite Element computations of the block surface is described in [5]. The following describes briefly the type of FE models required and some typical results.

6.6.1 Finite Element Models

Figure 10 illustrates a typical full cylinder block model of a four-cylinder engine. Typically, it is necessary to use a model of 1500 elements of the three or four node plate bending type. This gives rise to about 4000 total degrees of freedom with a typical 'front' size of 460. In order to ensure adequate accuracy of the natural frequencies of the cylinder block, the eigenvalue extraction must be conducted with a minimum of 200–300 master degrees of freedom. Thus prior to this calculation, an initial stiffness and mass matrix reduction must be conducted. (Where cylinder blocks can be regarded as symmetrical structures (e.g. front and back halves of the block), problem size may be reduced by modelling one half of the engine but running the model with both symmetric and anti-symmetric constraints. These runs provide independently the 'odd' and 'even' modes of vibration of the structure.)

6.6.2 Forced response analysis

The starting point for forced response analysis is, as in cylinder block stress analysis, the cylinder pressure diagram (Fig. 36). It is recommended that appropriately averaged data (i.e. in the frequency domain) are used to compute correct bearing and cylinder head loads etc. These load diagrams will be computed in the time domain and should be converted to frequency domain (Fig. 37) to allow a modal superimposition forced response analysis to be conducted [5].

6.6.3 Typical results

Figures 38 and 39 show typical predictions for the cylinder block of a four-cylinder diesel engine. The noise predictions have been conducted on the basis of the theory proposed in [6].

6.6.4 Extension to Analysis of Covers

The major noise sources of the engine, such as the sump, are forced directly by the cylinder block. If such covers have a significant stiffness/mass it is necessary to include that cover in the overall Finite Element model. This, of course, increases the total number of freedoms in the model. The sump model shown in Fig. 11 would typically add a further 1500 degrees of freedom to the problem size. Eigenvalues derived from similar sized reduced mass and stiffness matrices as used for the block above (section 6.6.1) probably lack the accuracy to provide correct frequencies and node shapes for the block and sump combined. However, increased numbers of master freedom yield a very large problem size in terms of solution time required and core space.

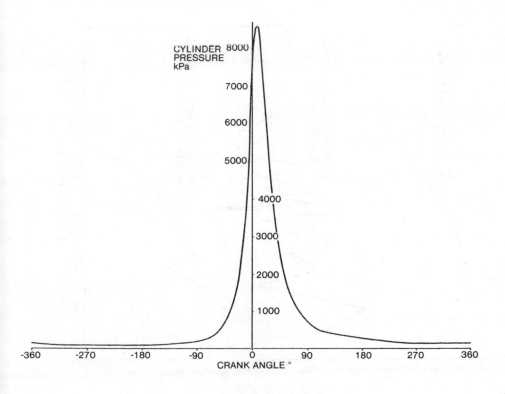

Fig. 36 – Cylinder pressure diagram.

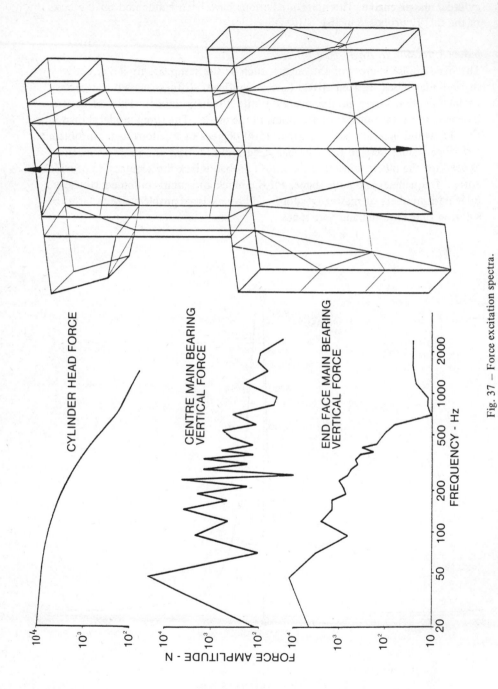

Fig. 37 – Force excitation spectra.

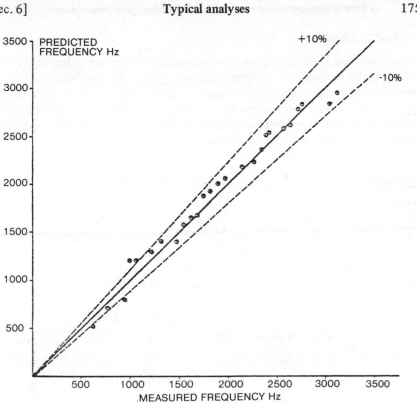

Fig. 38 – Natural frequency comparison.

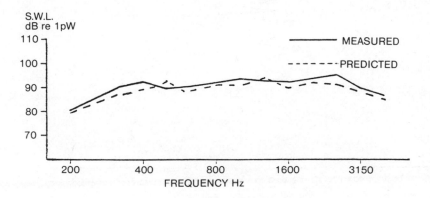

Fig. 39 – Block sound spectrum.

7. CONCLUSIONS

This chapter has attempted to describe an *approach* to the use of Finite Element models in the analysis of the complete engine system. In a number of instances where actual measurements have been reported, the engineers can see the accuracy with which FE analysis can be conducted. Obviously, all uses of FE analysis have not been discussed here, but it is anticipated that sufficient examples have been discussed to illustrate the value of this approach.

REFERENCES

[1] McAdams, W. H., *Heat Transmission*.
[2] Rohsenow and Choi, *Heat, Mass and Momentum Transfer*.
[3] Graff, W. J., 'Thermal Conductance across Metal Joints', *Machine Design*, September (1960), pp. 166–172.
[4] *PAFEC 70+/PAFEC 75 User Manuals*, PAFEC Ltd.
[5] Hawkins, M. G. and Southall, R., 'Analysis and Prediction of Engine Structure Vibration', *SAE 750832*.
[6] Hawkins, M. G. and O'Keeffe, J. M., 'A Method of Determining the Effect of Design Changes on Diesel Engine Noise, *CIMAC* (1975).

6

Crankshaft loading and bearing performance analysis

B. Law, Perkins Engines

---

1. INTRODUCTION

The calculation of bearing performance and crankshaft fatigue strength is becoming increasingly important in diesel engine design and development as more competitive engine specifications are sought. Main and large-end bearing performance and crankshaft loading predictions are required

(i) to optimize bottom end dimensions of new engines,
(ii) to aid in the solution of problems in engine development and
(iii) to provide loading for engine stress (finite element and experimental) and noise analyses.

The objective of this chapter is to describe the general methods used to predict crankshaft loading and bearing performance with some reference to the method of main bearing load calaculation developed by the writer.

2. BEARING ANALYSIS

Before dealing with specific cases (large-end bearing [section 3] and main bearings [section 4]) it is necessary to consider the established procedures used to assess dynamically loaded engine bearings.

A satisfactory journal bearing must satisfy the following two requirements:

(i) The bearing (and journal) must withstand the oil film pressure loading without mechanical (fatigue) failure.
(ii) An oil film of sufficient thickness to avoid seizure or rapid wear must be generated.

The methods usually employed in assessment of fatigue resistance and hydrodynamic performance are described in sections 2.1 and 2.2.

In addition, other factors such as material compatibility and hardness, oil feed mechanism, and oil filtration must be given consideration. These matters are discussed in [1].

2.1 Bearing Fatigue Strength

The calculation of oil film pressure in a flexible dynamically loaded bearing is a very complicated matter currently being addressed by finite element methods, It is apparent from a paper by Martin *et al.* [3] that a fatigue assessment based on oil film pressures (which were computed from a simplified rigid model) is also complex. To circumvent these difficulties the bearing specific load, defined below, is used to assess fatigue loading.

$$\text{specific load} = \frac{\text{bearing load}}{\text{bearing projected area}}$$
$$\text{(i.e. land length} \times \text{diameter)}$$

Given the applied loading diagram (as a function of crank angle) the maximum specific load can be computed. The maximum specific load is compared with empirically derived limits [3] provided by the bearing manufacturers. Specific load limits for common bearing materials are presented in [1]. The limits usually refer to plain bearings fed by a single oil hole (e.g. connecting rod large-end bearings in automotive diesels). For grooved bearings the limits should be factored by 0.75 to reflect the effect of the higher pressures generated in a grooved bearing compared to a plain bearing supporting the same load (see Fig. 1).

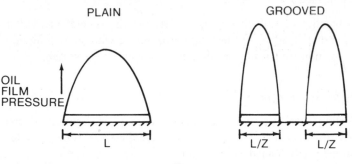

Fig. 1.

2.2 Hydrodynamic Assessment

The Hydrodynamic analysis, which involves calculation of the journal orbit diagram, permits an evaluation in respect of seizure and rapid wear. The Mobility Method created by Booker [4] is used extensively in the motor industry for orbit calculations and will be described in some detail (section 2.2.1). Other methods will be briefly mentioned (section 2.2.3).

2.2.1 Mobility Method

Inspection of Reynold's equation (which is pertinent to the analysis of thin lubrication films) reveals two mechanisms for generation of oil film pressure in cylindrical journal bearings. The first is due to the circumferential 'wedge' oil film shape resulting from journal eccentricity. The second mechanism, termed squeeze action, is pressure generation by journal centre motion with the bearing clearance space. It is a combination of 'squeeze' and 'wedge' pressure terms which dictates the journal trajectory in a dynamically loaded bearing.

The 'mobility' method is based upon force/velocity (mobility) calibrations of the bearing oil film. The calibrations (which may be derived analytically from a degenerated Reynold's equation) are used in conjunction with the following equation of motion to predict the journal centre velocities:

$$\begin{bmatrix} \dot{e}_x \\ \dot{e}_y \end{bmatrix} = \frac{4F(c/D)^2}{\mu LD} \begin{bmatrix} M_x \\ M_y \end{bmatrix} + \frac{\omega}{2} \begin{bmatrix} 0 & -1 \\ 1 & 0 \end{bmatrix} \begin{bmatrix} e_x \\ e_y \end{bmatrix} \tag{1}$$

The above equation is used to march out, in time, a solution for the journal centre position (e_x, e_y) using the following procedure.

(i) Set (e_x, e_y) (Fig. 2) to arbitrary initial values (at time $t = 0$). Use equation (1) to compute journal centre velocities relative to the bearing.

(ii) After a small time Δt, during which the crank turns $\Delta \theta$, the new journal centre positions may be evaluated as follows:

$$\begin{bmatrix} e_x \\ e_y \end{bmatrix}_{t + \Delta t} \approx \begin{bmatrix} e_x \\ e_y \end{bmatrix}_t + \Delta t \begin{bmatrix} \dot{e}_x \\ \dot{e}_y \end{bmatrix}_t \tag{2}$$

(iii) Obtain the force applied to the bearing at crank angle $\theta + \Delta \theta$ and re-evaluate equation (1) substituting

$$\begin{bmatrix} e_x \\ e_y \end{bmatrix}_{t + \Delta t}$$

to obtain

$$\begin{bmatrix} \dot{e}_x \\ \dot{e}_y \end{bmatrix}_{t + \Delta t}$$

(iv) Repeat step (ii) with t replaced by $t + \Delta t$ and then repeat step (iii).

During the marching process the transient due to the initial journal position decays and the solution becomes cyclically repetitive (after about one-and-a-half engine cycles).

Mobility components along and perpendicular to the load vector direction may be derived from a solution of Reynold's equation after making the Ocvirk

DYNAMICALLY LOADED BEARING ANALYSIS

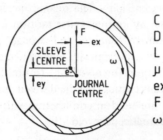

C   DIAMETRAL CLEARANCE
D   DIAMETER
L   LENGTH
μ   VISCOSITY
ex,ey ECCENTRICITY COMPONENTS
F   APPLIED LOAD
ω   SPEED

USING THE 'MOBILITY' CONCEPT RATES OF CHANGE OF ECCENTRICITY ARE
CALCULATED FROM :-

$$\begin{bmatrix} \dot{e}x \\ \dot{e}y \end{bmatrix} = 4\frac{F(C/D)^2}{\mu LD} \begin{bmatrix} Mx \\ My \end{bmatrix} + \frac{\omega}{2}\begin{bmatrix} 0 & -1 \\ 1 & 0 \end{bmatrix}\begin{bmatrix} ex \\ ey \end{bmatrix}$$

THE MOBILITIES ARE DERIVED FROM :-
   (i) A DEGENERATED REYNOLDS' EQUATION
OR (ii) CURVE FITS TO NUMERICAL SOLUTIONS (F.D./ F.E.) OF REYNOLDS' EQUATION

$$\begin{bmatrix} Mx \\ My \end{bmatrix} = f\,(ex, ey, L/D,)$$

Fig. 2 – Dynamically loaded bearing analysis.

short bearing asumption [4]. Mobility components have been derived by curve
fits to numerous finite difference solutions of Reynold's equation by Moes *et al.*
[5]. The former mobilities over-estimate the oil film stiffness; the latter mobilities
are more accurate.

The purpose of the above description is to convey a general solution philosophy for bearings; it is not a rigorous justification or theoretical description of
the mobility concept.

2.2.2 Interpretation of the Results
The 'mobility' method (and other similar methods) has been in use within the
motor industry for a number of years and there is considerable experience [6] in
the interpretation of the locus diagram.

General acceptance levels for high-speed diesel engine bearings, in terms of minimum oil film thickness (based on short bearing assumptions), are:

0.75 μm   Grooved bearings
1.2 μm   Plain bearings

The shape of the locus diagram is analysed when assessing cavitation erosion [7] or wear problems [8].

2.2.3 Other methods

In recent years, considerable advances in the understanding of journal bearing operation have resulted from finite element and finite difference solutions of Reynold's equation [9,5]. These approaches do not embody the restrictive assumptions made in analytic solutions of Reynold's equation. The methods, which are currently being developed for analysis of engine bearings (dynamically loaded and flexible), will not be discussed further in this chapter.

3. CONNECTING ROD AND LARGE-END BEARING

Calculation of the loading on this bearing is a prerequisite to the analysis described in section 2. The loading is readily derived from the data presented in Fig. 3 (component masses and dimensions and the cylinder pressure diagram) as a function of crank angle. The acceleration forces are best derived by analysis of the crank slider velocity and acceleration diagrams (see Swanson [10], p. 114).

Alternatively, the calculations can be based on the approximate expression for force on the connecting rod small end due to acceleration of the reciprocating masses.

$$F \approx -mr\omega^2 \left(\cos\theta + \frac{r}{l} \cos2\theta \right)$$

where

F is the force on the connecting rod
r is the crank throw
l is the connecting rod length
m is the reciprocating mass (piston + small end of rod)
ω is the crankshaft rotation speed.

Summation of acceleration and gas pressure forces provides the load diagram for the large-end bearing.

Bearing load and orbit diagrams computed using the data in Fig. 3 are presented in Fig. 4.

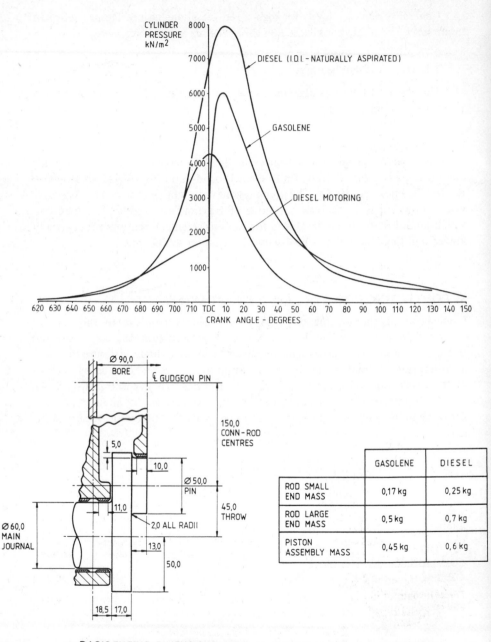

Fig. 3 – Example data for large-end bearing analysis.

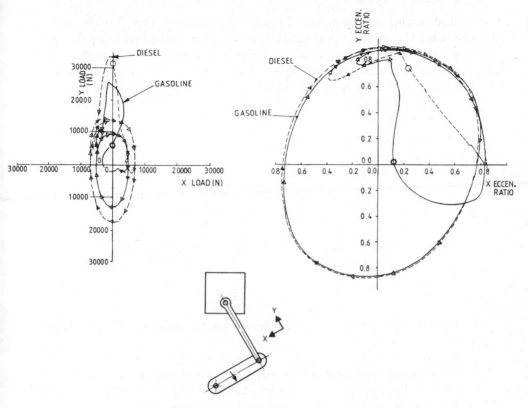

POLAR LOAD AND ECCENTRICITY RATIO FOR
LARGE END BEARING.
FULL LOAD – ENGINE SPEED-4500 RPM.

	Full power 2500 rpm		Full power 4500 rpm		Zero power 5750 rpm	
	Min. oil film thickness* μm	Max. specific load kN/m²	Min. oil film thickness* μm	Max. specific load kN/m²	Min. oil film thickness* μm	Max. specific load kN/m²
Gasoline	2.23	<u>34.4</u>	1.46	25.5	<u>1.37</u>	20.9
Diesel loading and components	1.58	<u>45.5</u>	1.28	33.4	<u>1.15</u>	28.0

Summary of large-end bearing performance
*('mobility' with Ocvirk assumptions used)

Fig. 4 – Gasoline/diesel – large-end bearing performance.

4. MAIN BEARINGS

Main bearing load and orbit diagrams, equivalent to those for the large-end bearing, are required to assess these bearings. Main bearing load calculation is not a simple matter. The crankshaft is a continuous flexible structure supported by a number of main journal bearings which are carried in a flexible engine structure. Known loadings are applied to the crankshaft and engine structures.

The system is indeterminate and therefore both equilibrium and compatibility equations have to be satisfied. Modelling of:

 (i) crankshaft stiffness,
 (ii) engine stiffness, and
(iii) main journal bearing operation,

is necessary to analyse the system properly.

The statistically determinate method, which is widely used in the motor industry for calculation of main bearing load, is described in section 4.1. Alternative load calculation methods are described in section 4.2 as an introduction to the procedure developed by the writer which is outlined in section 4.3.

4.1 Statistically Determinate Main Bearing Load Calculation

The crankshaft system is analysed using a determinate method by assuming that each span of the crankshaft between main bearings is simply supported. Consequently, forces and moments cannot be transferred across crankshaft main journals. The assumption corresponds to a crankshaft with hinge joints at the main journal centres; see Fig. 5.

The load components $(F_{x_L}, F_{y_L}$ and $F_{x_R}, F_{y_R})$ acting on the main bearing due to force components P_x and P_y acting on a crankthrow (see Fig. 6) are:

$$F_{x_L} = P_x L_R/(L_L+L_R), F_{Y_L} = P_Y L_R/(L_L+L_R)$$

$$F_{x_R} = P_x L_L/(L_L+L_R), F_{Y_R} = P_Y L_L/(L_L+L_R)$$

where L_L and L_R are the axial distances from the force P to the left and right bearings.

The main bearing load is calculated by distributing the crankpin and centrifugal loadings according to the equations given above. Bearing load components from adjacent crankthrows must be added to yield the total load on intermediate main bearings. The general principle is indicated in Fig. 5 for one co-ordinate direction. The crankpin forces P_2, P_3, P_4 are derived by consideration of the phase angle differences implied by the engine firing order. (If the firing order is 1–3–4–2 and the crankpin force at crank angle α is F_α, then the forces P_3, P_4 and P_2 (Fig. 5) are $F_{(\alpha - 180)}, F_{(\alpha - 360)}$ and $F_{(\alpha - 540)}$.

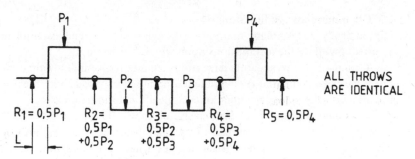

STATICALLY DETERMINATE METHOD

Main bearing no.	Engine loading and components	Full power 2500 rpm		Full power 4500 rpm		Zero power 5750 rpm	
		Min. oil film thickness μm	Max. specific load kN/m²	Min. oil film thickness μm	Max. specific load kN/m²	Min. oil film thickness μm	Max. specific load kN/m²
1	Gasoline	1.70	11.7	1.46	9.27	1.22	15.1
	Diesel	1.51	16.0	1.36	11.0	1.05	18.2
2	Gasoline	1.95	14.1	2.26	13.2	7.32	2.4
	Diesel	1.46	19.4	1.71	18.1	2.64	7.1
3	Gasoline	1.33	8.94	1.23	15.0	0.81	30.3
	Diesel	1.23	12.8	1.06	19.5	0.83	30.1

Summary of main bearing performance
with gasoline, and with diesel, cylinder pressures and components

Fig. 5 – Gasoline/diesel – main bearing performance.

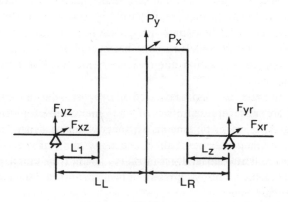

Fig. 6.

4.2 Other methods – a literature review

The statically determinate method, although easily applied, is fundamentally incorrect owing to the assumptions upon which it is based.

A number of investigators have proposed alternative indeterminate methods and some have compared their predicted main bearing loads with measurements and also with the loads calculated on a statically determinate basis.

Gross and Hussman [11] have measured the vertical component of the main bearing load acting on the cap half of the main bearings of two in-line (Daimler–Benz and MAN) six-cylinder diesel engines. They compared the measurements with the predictions of the statically determinate method and two indeterminate methods. In both of the indeterminate methods the crankshaft was idealized as a continuous, straight, circular bar having a diameter of about 65% of the main journal diameter. In the first indeterminate procedure the main bearing loads were calculated with the crankshaft model rigidly supported (in translation) at the main bearings. In the second calculation method the crankshaft model was supported by linear, grounded springs (representing the combined stiffness of the oil films and bearing housings).

Measured loads are presented for the engine at idling conditions. The results obtained from the first indeterminate procedure (rigid supports) are in poor agreement with the measurements (discrepancies of the order of 50% in peak bearing loads); the statically determinate predictions were found to be in closer agreement with the measurements.

In most cases the second indeterminate method produced results in closer agreement with measurements (discrepancies of the order of 25%) than the results of the statically determinate analysis. Gross and Hussman found that the support spring stiffness (used in the second indeterminate procedure) corresponding to the peak bearing load divided by the bearing clearance resulted in the best comparison with measurements.

Brach [12] and Porter [13] have presented indeterminate main bearing computation procedures involving cranked beam models. The cranked shafts were supported via linear springs, representing bearing and oil film elasticities. Brach quotes differences in the main bearing loads of up to 20% between cranked and uniform (as per Gross and Hussman) beam crankshaft representations. In both cases the theoretical method due to Timoshenko [14] was used to create the cranked-shaft model.

The indeterminate methods reviewed above use over-simplified oil film models which are not representative of the operation of journal bearings. Journal bearings have an instantaneous stiffness which is dependent upon the journal eccentricity, and in general, the eccentricity and force vectors are not in the same direction. The load and eccentricity diagrams in Fig. 4 readily show that the journal bearing is poorly represented by a linear spring since the diagrams do not have the same shape.

Von Schnurbein [15] was the first to incorporate a representative journal bearing model into the indeterminate system. Von Schnurbein sequentially solved the structural equations of a uniform circular beam, representing the crankshaft, with the hydrodynamic equations governing the motion of the journals within the bearing.

The procedure operates as follows:

At each step in the computation scheme the structural equations are solved (using a transfer matrix method) to produce main bearing loads given the journal eccentricities within each bearing. The hydrodynamic equations are then applied to predict the new journal of eccentricities after a time increment for use in the next calculation step (a procedure similar to that described in section 2.2). The successful fusion of the hydrodynamic and structural models represented a significant advance in main bearing load calculation. Von Schnurbein compared the predictions of his method to Gross and Hussman's predictions and measurements. The comparisons presented, all at idling conditions, are encouraging, with only 10% discrepancy between Von Schnurbein's predictions and measurements at the peak load conditions (there is greater discrepancy at other points in the cycle).

In a Cornell University thesis, Stickler [16] has developed a calculation method for journal bearings in indeterminate systems. The governing structural and hydrodynamic equations are solved sequentially by the same marching technique used by Von Schnurbein. Stickler has used the 'mobility' formulation to model the journal bearings and a finite element approach for the crankshaft. The finite element model of the crankshaft, which comprises beam elements, is used to compute sets of influence coefficients. These allow a structural treatment of the crankshaft which is both computationally efficient and conceptually appealing — far more so than the transfer matrix methods or the rather lengthy beam theory treatments used by other investigators.

4.3 Main bearing load calculation — an indeterminate method

Vast increases in digital computing power and availability of the finite element method of structural analysis permitted the development of an advanced load calculation procedure.

The procedure is based on a sequential solution of the structural equations (written in terms of crankshaft and engine influence coefficients) and the hydrodynamic equations governing the journal motion (using the 'mobility' concept).

The influence coefficients are derived by finite element methods as described in sections 4.3.1 and 4.3.3. The operation of the solution strategy is presented in section 4.3.2.

4.3.1 Crankshaft influence coefficients

The structural equation for the crankshaft is

$$[f_j] = [T_c] [f_p] + [k_c] [u_j] + [c] w^2 \tag{3}$$

where:

[f_j]   is a column vector of main journal (reaction) forces.

[f_p]   is a column vector of applied crankpin loads.

[u_j]   is a column vector of main journal displacements.

[c]   is a column vector of main journal (reaction) forces due to crankshaft rotation at 1 rad/s with zero journal displacements and crankpin forces.

[T_c]   is a matrix of transmissibility influence coefficienets. Each column comprises the journal reaction forces produced by a particular unit crankpin load (see Fig. 7) in the absence of centrifugal load and journal displacement.

[k_c]   is a matrix of stiffness influence coefficients. Each column comprises the journal forces required to generate a particular unit journal displacements at the other journal degrees of freedom and zero crankpin and centrifugal loading.

w is the crankshaft rotation speed (rad/s)

The above equation permits efficient computation of journal forces due to prescribed crankpin loads, journal displacements and crankshaft rotation speed.

The influence coefficients are calculated from the stiffness matrix and centrifugal load vector (relative to the appropriate journal and crankpin degrees of freedom) derived by finite element methods (see appendix).

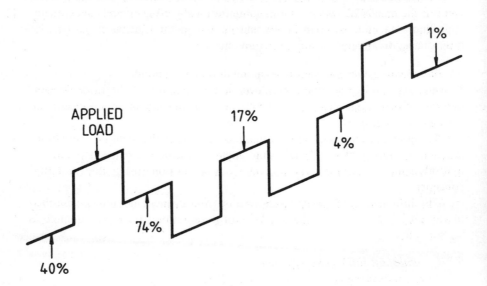

Fig. 7 – Illustration of transmissibility influence coefficients.

4.3.2 Solution Procedure

To simplify the explanation of the procedure, the main bearings will be assumed rigid. The vector $[u_j]$ in equation (3) therefore, comprises the journal eccentricities, $[e]$ (e_x and e_y components for each journal). Solutions for main bearing loads (the negative of $[f_j]$ and eccentricities are marched out using the following procdure.

(i) Set $[u_j]$ arbitrarily to zero and use equation (3) to evaluate the main bearing forces $(-[f_j])$.

(ii) Calculate the journal centre velocities due to the main bearing forces (equation (1)).

calculation loop

(iii) After a short time (typically that corresponding to 0.5° crank angle) the position of each journal centre is given by equation (2).

(iv) Calculate the main bearing forces (equation (3)) using the current journal centre positions and crankpin forces appropriate to the (incremented) crank angle.

The transient due to the initial conditions rapidly decays to yield a cyclically repetitive solution for main bearing loads and journal eccentricities after about one-and-half engine cycles.

The above procedure may be extended to include bearing flexibility. An equation similar to (3) is written for the full engine. When this equation is combined with equation (3) (by applying force and displacement compatibility equations), an expression for main bearing load in terms of engine forces (gas pressure and side forces from the pistons), journal eccentricities, crankpin forces and centrifugal loading is produced which may be substituted for equation (3) in the above procedure.

4.3.3 Derivation of the influence coefficients

The influence coefficients are derived from the crankshaft and engine stiffness matrices produced by finite element methods. The engine substructured model discussed in Chapter 5 provides the required engine stiffness matrix relative to bearing, cylinder and engine mounting degrees of freedom.

Three-dimensional solid elements must be used in the crankshaft model to provide an accurate stiffness matrix and centrifugal load vector relative to crankpin and main journal degrees of freedom. (Beam theory stiffness models are grossly inaccurate in crankshafts with overlap; see Hildrew [17].) It is necessary to substructure the crankshaft to avoid very long computer runs. A convenient substructure is a half crankthrow. A symmetric half crankthrow substructure may be generated by summing the results from a quarter throw model analysed with symmetric and antisymmetric constraint conditions on the plane of symmetry (Fig. 8). The entire crankshaft model is built up by merging the substructures after appropriate orientation manipulations have been carried out.

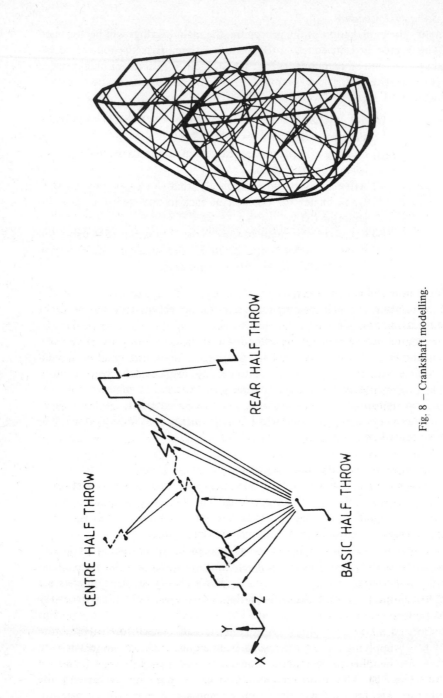

Fig. 8 – Crankshaft modelling.

After the assembly process the crankshaft stiffness matrix and centrifugal load vector may be extracted relative to the appropriate degrees of freedom (see appendix).

5. CRANKSHAFT LOADING AND STRESSING

This section deals with analysis of crankshaft loading between the extreme main journals. The recognized vulnerable areas in respect of fatigue failure are:

(i) The fillet radii between the crank journals and the webs (see Fig. 9, AA).
(ii) The crankpin surfaces at the intersection with oil drillings.

The stresses in the fillet radii are primarily due to bending loads in the plane of the crankthrow. The crankpin surface stresses are due to torsional loading from both vibration and the internal torques produced within the crankshaft.

The crankshaft analysis must include assessments of the above vulnerabilities which will be discussed in section 5.1 and 5.2.

5.1 Vulnerability in the fillet radius

Three methods may be used to assess the crankshaft integrity. All methods require calculation of the loading applied to each crankthrow (in particular the bending moment in the plane of the crankthrow at the centre of the crankwebs). The calculation of crankthrow loading is trivial when statically determinate assumptions are applied. For example, with reference to Fig. 6, the crankweb bending moments due to load P are $F_{y_L} \times L_1$ and $F_{y_R} \times L_2$.

Loading transmission along the crankshaft is permitted in indeterminate analyses. The loading at any position in the crankshaft (generally three forces and three moments) is the summed effect of all the forces acting on the front or rear section of the crankshaft from the point of interest. This computation canot be carried out until the main bearing load has been calculated.

The first method is based on stress calculation using detailed finite element models. This method uses the loading (forces and moments) at the boundary faces of each of the half crankthrow substructures. The stresses at a number of points in the fillet radii are derived by scaling the fillet stresses due to unit values of the loading on the substructure faces by the loadings at each crank angle. The stresses due to unit loadings are derived by the finite element method. A Goodman/Gerber fatigue reserve factor is derived for each of the points in the fillet. This method requires considerable computer resources to process the detailed finite element models.

The second method of fatigue assessment is based on fatigue testing of the crankshaft. By applying a sinusoidal bending moment of known amplitude to a crankweb, the bending fatigue strength at one of the fillet radii may be determined (see Fig. 10). A fatigue diagram is constructed on the basis of the rig test bending moments for the fillet which has been tested. This often involves

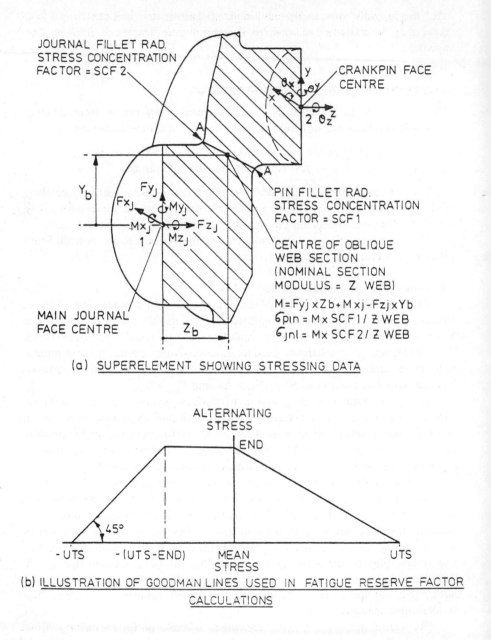

JOURNAL FILLET RAD.
STRESS CONCENTRATION
FACTOR = SCF 2

CRANKPIN FACE
CENTRE

PIN FILLET RAD.
STRESS CONCENTRATION
FACTOR = SCF 1

CENTRE OF OBLIQUE
WEB SECTION
(NOMINAL SECTION
MODULUS = Z WEB)

$M = Fyj \times Zb + Mxj - Fzj \times Yb$
$\sigma_{pin} = M \times SCF1 / Z\ WEB$
$\sigma_{jnl} = M \times SCF2 / Z\ WEB$

MAIN JOURNAL
FACE CENTRE

(a) SUPERELEMENT SHOWING STRESSING DATA

ALTERNATING
STRESS

END

45°

- UTS   -(UTS-END)   MEAN   UTS
STRESS

(b) ILLUSTRATION OF GOODMAN LINES USED IN FATIGUE RESERVE FACTOR
CALCULATIONS

Fig. 9 – Crankshaft stressing and fatigue calculation data.

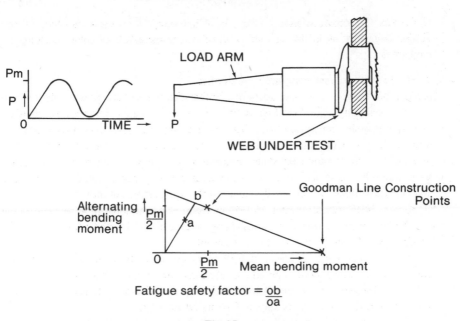

Fatigue safety factor $= \dfrac{ob}{oa}$

Fig. 10.

either specifying the endurance ratio of the material (endurance limit/ultimate strength) or calculating the bending moment which corresponds to the material ultimate tensile stress in the fillet. The mean and alternating bending moments applied to the crankweb during the engine cycle are obtained from the minimum and maximum calculated moments. These mean and alternating moments are plotted onto the Goodman diagram (see 'a' in Fig. 10) and the fatigue safety factor (FSF) is obtained. This method is far simpler to apply than the first method but it neglects the effect of other loadings and requires rig test results as data.

The final method of fatigue assessment is based on fillet bending stresses given by empirically derived stress concentration factors. A number of empirical correlations have been reported [18, 19]. All of these are based on a nominal stress defined by the crankweb bending moment divided by the section modulus of a reference section. The reference section is usually a section across the crankweb. The fillet stress is the nominal stress multiplied by the stress concentration factor derived by either graphical presentation or regression fits of the experimental results. Stress concentration factors are usually in the range 3 to 5 and are functions of the crankthrow dimensions (web width, crankpin diameter, overlap, fillet radius, web thickness, collar dimensions). Using the mean and alternating fillet stresses and the fatigue properties at the fillet region of the crankshaft (which are often derived by analysis of the rig bending moments described under the first method), the fatigue reserve factor may be calculated from the Goodman diagram (Fig. 9). This method is easily applied at the design stage but, as in the second method, the stresses due to other loadings are neglected.

5.1.1 Acceptance levels/material properties

Acceptable fatigue reserve factors for crankshafts depend strongly on the method used to calculate the bending moments. The statically determinate method provides grossly inaccurate bending moment predictions. This was illustrated during the experimental verification of the indeterminate method presented in section 4.3. The crankweb strain measured in a running engine was compared to predicted strains according to loadings derived by the statically determinate and indeterminate methods. Typical results are presented in Fig. 11. General agreement with measurements of 10% in strain range was found in respect of the indeterminate predictions; the strain range predicted by the statically determinate method was 10–80% greater than measured.

Acceptance levels in terms of FSF appropriate to the statically determinate method vary with engine type and crankshaft and engine stiffnesses in the range 0.9–1.3. Judgement of crankshaft integrity based on statically determinate calculations relies heavily on service history knowledge.

Treatments may be applied to crankshafts to enhance the fatigue resistance in the fillet region. As almost all these treatments induce residual stresses in the fillets the experimental derivation of bending strength cannot be based on a single type of rig loading (e.g. one-way bending). Typically, the endurance limit in the ground fillet of a forged steel crankshaft is 370–420 N/mm^2 for a UTS of 850 N/mm^2. Approximate increases in the strength of this crankshaft in terms of a one-way bending loading (0 to maximum moment) are:

> 30–40% due to tufftriding
> >   60% due to nitriding (60 hours)
> 30–50% due to fillet rolling

5.2 Crankpin/oil drilling vulnerability

Similar stress analysis methods to those discussed in section 5.1 are applicable to this region. The (non-vibratory) torsion load at the crankpins due to engine operation is calculated using the general method described in 5.1.

Torsional vibration analysis is well documented (e.g. [20]). In most TV analyses, the crankshaft and driven equipment are represented by a mathematical model comprising concentrated inertias (e.g. at the flywheel, crankpins and crankshaft pulley) connected by torsion springs. The natural frequencies, mode shapes and torques in this model are calculated by Holzer's tabulation. The torques and mode shapes correspond to an arbitray unit amplitude, usually at the crank pulley inertia. A dynamic magnifier (typical values must be determined by amplitude measurements) is required in the calculation of the forced vibration amplitudes and torques.

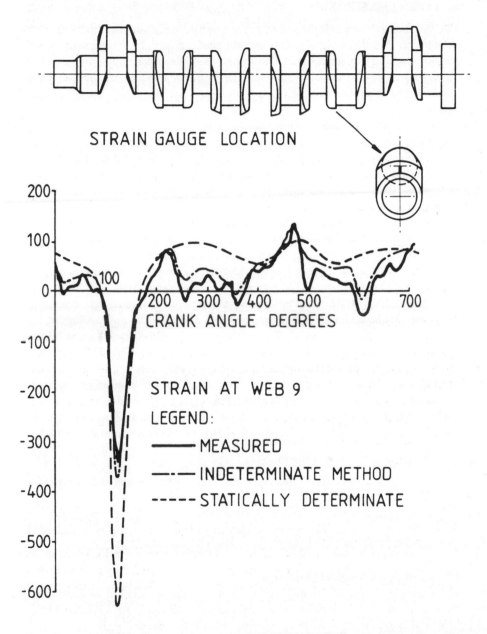

Fig. 11.

6. TYPICAL RESULTS

The first example is a comparison of bearing performance and crankshaft strength predictions by the statically determinate method and by the indeterminate method discussed in section 4.3. The second example concerns the prediction of crankshaft motion at the seal locations. The statically determinate method is not applicable to this prediction.

6.1 Determinate and Indeterminate predictions

The comparison of the statically determinate method and the indeterminate method discussed in section 4 is centred on the gasoline engine data (Fig. 3).

The half throw crankshaft model and the influence coefficients are presented in Fig. 12. A suitable four-cylinder engine finite element model was used to derive the engine influence coefficients.

The main bearing performance summary, presented in Fig. 13 (a), shows significant differences between the calculation methods in a number of results. The critical results (underlined) are, however, quite similar.

The fatigue safety factors at each fillet radius (Fig. 13 (b)) were calculated using an endurance stress of 370 N/mm² and an ultimate stress of 850 N/mm² (EN19 steel, forged). All the fatigue safety factors for the journal (and crankpin) fillets are identical in each operating condition according to the loadings of the statistically determinate method. The results are identical because all crankthrows are identical and each throw is assumed to be simply supported. The fatigue safety factors by the indeterminate method are considerably different to those produced by the determinate method. The indeterminate predictions indicate the journal fillets of webs 3 and 4 to be significantly more vulnerable than the other fillets. The critical operating speed for crankshaft strength is 2500 rpm by the statically determinate method and 4500 rpm by the indeterminate method.

6.1.1 Conclusion

The comparison clearly illustrates a general finding that crankshaft loading is far more sensitive to calculation method than bearing performance.

A statically determinate approach (i) incorrectly predicts the magnitude of the crankshaft loading, (ii) predicts (incorrectly) equal vulnerability to failure at all crankwebs, (iii) does not identify the critical engine operating condition.

6.2 Assessment of crankshaft oil seals

The example presented in this section concerns the prediction of crankshaft motion at the oil seal locations. The crankshaft translation motion relative to the front seal is calculated using the geometric relationship shown below

$$\begin{bmatrix} u_x \\ u_y \end{bmatrix} = \begin{bmatrix} e_x \\ e_y \end{bmatrix} + \begin{bmatrix} u_{b_x} \\ u_{b_y} \end{bmatrix} + \begin{bmatrix} 0 & -Z_s \\ Z_s & 0 \end{bmatrix} \begin{bmatrix} \theta_x \\ \theta_y \end{bmatrix}$$

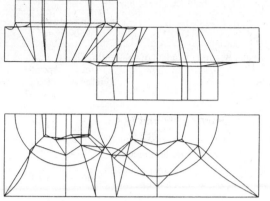

Fig. 12 — Influence coefficients and superelement FE model.

32 ELEMENT IDEALISATION

JOURNAL REACTIONS IN EACH COLUMN ARE DUE TO THE UNIT CRANKPIN LOAD (IN DIRECTION X,Y,YMETA X OR THETA Y) GIVEN IN THE COLUMN HEADING. THE JOURNAL REACTIONS MAINTAIN ZERO DISPLACEMENT IN THE REACTION DIRECTIONS.

PIN NO.—	1 X	1 Y	2 X	2 Y	3 X	3 Y	4 X	4 Y
1 X	-2398	0	-322	0	8	0	3	0
1 Y	0	-3814	0	-826	0	259	0	101
2 X	-10235	0	5556	0	-104	0	-37	0
2 Y	0	-7733	0	5726	0	-1446	0	-562
3 X	399	0	2594	0	2594	0	399	0
3 Y	0	2007	0	6287	0	6287	0	2007
4 X	4500	0	4433	0	10093	0	-5698	0
4 Y	0	-562	0	-1446	0	5726	0	-7733
5 X	-2266	0	-2260	0	-2591	0	-4667	0
5 Y	0	101	0	259	0	-826	0	-3814

*************** MATRIX STIFFNESS   (MULTIPLIED BY .1)

JOURNAL REACTIONS IN EACH COLUMN ARE DUE TO THE UNIT JOURNAL DISPLACEMENT (IN DIRECTION X,Y,YMETA X OR THETA Y) GIVEN IN THE COLUMN HEADING. THE JOURNAL REACTIONS MAINTAIN ZERO DISPLACEMENT IN THE REACTION DIRECTIONS OTHER THAN THE UNIT DISPLACEMENT. THE COLUMN HEADING REFERS TO.

JNL. NO.—	1 X	1 Y	2 X	2 Y	3 X	3 Y	4 X	4 Y	5 X	5 Y
1 X	4390	0	-9195	0	5259	0	-493	0	39	0
1 Y	0	3479	0	-8014	0	5886	0	-1647	0	295
2 X	-9195	0	23649	0	-20206	0	6245	0	-493	0
2 Y	0	-8014	0	21914	0	-21432	0	9179	0	-1647
3 X	5259	0	-20206	0	29895	0	-20207	0	5259	0
3 Y	0	5886	0	-21432	0	31094	0	-21432	0	5886
4 X	-493	0	6245	0	-20207	0	23649	0	-9195	0
4 Y	0	-1647	0	9179	0	-21432	0	21914	0	-8014
5 X	39	0	-493	0	5259	0	-9195	0	4390	0
5 Y	0	295	0	-1647	0	5886	0	-8014	0	3479

*************** MATRIX CENTRIFUGAL LOAD(MULTIPLIED BY 1.E+05)

JOURNAL REACTIONS DUE TO A ROTATION SPEED OF 1 RAD/S

JNL. DIRECTION	
1 X	0
1 Y	-2036
2 X	0
2 Y	-1760
3 X	0
3 Y	7592
4 X	0
4 Y	-1760
5 X	0
5 Y	-2036

Bearing	Calculation type	Full power 2500 rpm		Full power 4500 rpm		Zero power 5750 rpm	
		Min. oil film thickness μm	Max. specific load kN/m²	Min. oil film thickness μm	Max. specific load kN/m²	Min. oil film thickness μm	Max. specific load kN/m²
Main 1	Indeterminate	1.82	10.7	1.48	9.05	1.20	14.82
	Statically determinate	1.70	11.7	1.46	9.27	1.22	15.1
Main 2	Indeterminate	1.66	16.4	2.12	15.2	8.00	1.75
	Statically determinate	1.95	14.1	2.26	13.2	7.32	2.40
Main 3	Indeterminate	1.42	8.70	1.07	15.5	0.79	30.0
	Statically determinate	1.33	8.94	1.23	15.0	0.81	30.3

(a) Main bearing performance summary

Crank Web	Calculation type	Full power 2500 rpm		Full power 4500 rpm		Zero power 5750 rpm	
		Journal fillet	Crankpin fillet	Journal fillet	Crankpin fillet	Journal fillet	Crankpin fillet
1	Indeterminate	2.73	2.12	2.22	2.50	2.19	6.87
	Statically determinate	1.76	1.29	1.57	1.63	1.59	4.66
2	Indeterminate	3.51	2.75	2.23	2.63	1.82	3.46
	Statically determinate	1.76	1.29	1.57	1.63	1.59	4.66
3	Indeterminate	2.37	2.15	1.64	1.73	2.59	5.69
	Statically determinate	1.76	1.29	1.57	1.63	1.59	4.66
4	Indeterminate	2.39	2.13	1.62	2.12	2.55	10.1
	Statically determinate	1.76	1.29	1.59	1.63	1.59	4.66

(b) Crankshaft strength summary

Fig. 13 – Analysis of the gasoline engine (data Fig. 3).

where

u_x, u_y	are the displacements at the seal location
e_x, e_y	are the components of No. 1 journal eccentricity
u_{b_x}, u_{b_y}	are the components of No. 1 bearing displacement
θ_x, θ_y	are the slope components of No. 1 journal about the Cartesian axes
Z_s	is the axial distance between No. 1 journal centre and the seal location.

It is assumed that the seal housing does not displace and that the section of the crankshaft between No. 1 journal and the seal location does not bend.

The diagrams in Fig. 14 show the displacement of the crankshaft centreline at the seal location throughout the engine cycle in a typical four-cylinder high-speed diesel engine. The range of crankshaft movement is about 80% greater at 3600 rpm than at 1600 rpm.

7. CONCLUSIONS

The intention of this chapter is to provide a general view of design methods for crankshafts and for bearings. It is hoped that the following two conclusions will be drawn from the material presented.

(i) The validity of both bearing and crankshaft assessments is dependent on the accuracy of the bottom-end loading predictions.

(ii) The maximum accuracy of engine noise and stress analyses is limited by the accuracy of the main bearing loads which are used as input data in these analyses.

APPENDIX

Derivation of Crankshaft Influence Coefficients

Without loss of generality, the force—displacement equations of the crankshaft may be grouped as follows:

$$
\begin{bmatrix}
[f_j] + [f_{j_c}] \\
[f_p] + [f_{p_c}]
\end{bmatrix}
=
\begin{bmatrix}
S_{11} & S_{12} \\
S_{21} & S_{22}
\end{bmatrix}
\begin{bmatrix}
[u_j] \\
[u_p]
\end{bmatrix}
$$

where

$[S_{11}]$, $[S_{12}]$, $[S_{21}]$ and $[S_{22}]$ are partitions of the crankshaft stiffness matrix implied by the grouping of the equations

$[f_{j_c}]$, $[f_{p_c}]$, are the centrifugal force at the crankpin and main journal degrees of freedom calculated by finite element analysis

$[u_p]$ are the displacements at the crankpin degrees of freedom.

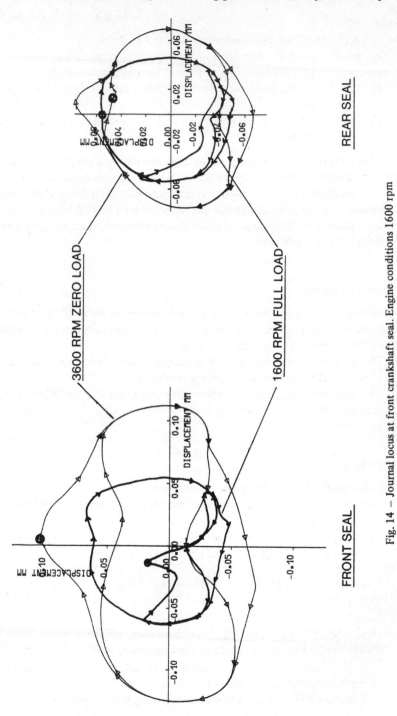

Fig. 14 — Journal locus at front crankshaft seal. Engine conditions 1600 rpm
max BMEP and 3600 rpm zero BMEP.

By expansion of the partitioned equations and elimination of $[u_p]$ the following expressions are obtained for the influence coefficients

$$[K_c] = [S_{11}] - [S_{12}] [S_{22}]^{-1} [S_{21}]$$
$$[T_c] = [S_{12}] [S_{22}]^{-1}$$
$$[C] = [T_c] [f_{p_c}] - [f_{j_c}] .$$

REFERENCES

[1] Neale, M. J., *Tribology Handbook*, Butterworth, London, 1973.

[2] Pratt, G. C., 'Materials for Plain Bearings', *Metallurgical Review* (1973), Institute of Metals (Glacier Co. Ltd. reference BM 403/73).

[3] Martin, F. A., Garner, D. R., Adams, D. R., 'Hydrodynamic Aspects of Fatigue in Plain Journal Bearings' (1978) Glacier Co. Ltd. reference LB 432/78.

[4] Booker, J. F., 'Dynamically Loaded Journal Bearings: Numerical Application of the Mobility Method', *I. Lub. Tech.*, **93**, No. 2 (1971).

[5] Moes, H., Childs, D. and von Leeuiven, H., 'Journal Bearing Impedance, Description for Rotodynamic Applications', *ASME paper 76 − Lub − 24* (1977).

[6] Booker, J. F., 'Design of Dynamically Loaded Journal Bearings', *Fundamentals of the Design of Fluid Film Bearings*, ASME (1979).

[7] Dunning, S. W., 'A Study of Cavitation Erosion under Conditions of Hydrodynamic Lubrication, *Ph.D thesis* Univ. of Leeds, 1980.

[8] Ross, J. M., 'Bearing Orbit Analysis', *Machine Design*, October (1971).

[9] Stafford, A., Henshell, R. D. and Dudley, B. R., 'Finite Element Analysis of Problems in Elasto hydrodynamic Lubrication', *5th Leeds Lyon Symposium*, Trans. ASME (1979), 406.

[10] Swanson, S. A. V., *Engineering Dynamics*, English Universities Press, London, 1963.

[11] Gross, W. and Hussman, A. W., 'Forces in the Main Bearings of Multicylinder Engines', *Trans. S.A.E.*, paper 660756 (1966).

[12] Brach, S., 'Reactions, Elasticity and Displacements of Bearings in Crankshaft Calculations', *MTZ*, No. 4 (1968), pp. 143−145.

[13] Porter, F. P., 'Crankshaft Stress Analysis and Bearing Load-carrying Capacity', *ASME paper* (68−D9P−8) (1968).

[14] Timoshenko, S., 'The Bending and Torsion of Multi-throw Crankshafts on Many Supports', *Trans. ASME*, **45** (1923), p. 449.

[15] Von Schnurbein, E., 'A New Method of Calculating Plain Bearings of Statistically Indeterminate Crankshafts', *Trans. ASME*, **79**, paper 700716 (1970).

[16] Stickler, A. C., 'Calculations of Bearing Performance in Indeterminate Systems', *Ph.D. thesis*, Cornell University, 1974.

[17] Hildrew, B., 'The Stress Analysis of Crankshafts', *50th Thomas Lowe-Gray Lecture, Proc. I. Mech. E.,* **192**, No. 31, (1978), pp. 30–37.

[18] Arai, J., 'The Bending Stress Concentration Factor of Solid Crankshaft', *Bulletin of JSME,* **8**, No. 31 (1965).

[19] Lang, O. R., 'Stress Concentration Factors in Crankshafts', *MTZ,* **29**, No. 3 (1968), pp. 91–95.

[20] Nestorides, E. J., *A Handbook on Torsional Vibration,* BICERA, Cambridge University Press, 1958.

7

Combustion noise in automotive diesel engines

M. F. Russell, Lucas Industries Noise Centre

1. INTRODUCTION

Noise made by road vehicles has become a serious nuisance to those living near urban main roads, industrial areas and motorways. The increasing use of diesel engines in private cars and light vans is causing greater general awareness of the noise they can produce. It is particularly noticeable in urban areas where noise from goods vehicles stands out from the background of traffic noise and community noise, by reason of 'combustion knock' and low frequency noise from the air intake and exhaust. Near an industrial area, the noise of trucks starting up and manoeuvring can be a nuisance, especially at night and in the early morning. Coupled with this greater awareness of vehicle noise, there are factors which are actually causing an increase in noise levels from engines; for example, the 'combustion knock' from diesel engines will become considerably more pronounced as the cetane number of diesel fuel drops significantly in the future, owing to changes in the mix of crude oils used to supply the fuel.

Legislation has been enacted to limit the noise emitted by new vehicles in 14 countries and these limits are due to be reduced by 3 dB(A) for most classes of vehicle in 1983/84. Proposals for eventual reductions in noise level, varying from 7 dB(A) below the present limits for cars to 11 dB(A) below the present limits for the largest trucks and buses, are being discussed. The proposed first date of introduction of lower noise limits is 1985 for European countries [1]. There are other proposals to modify the test procedure which tend to increase the noise level emitted by heavy goods vehicles as they undergo their type approval tests, thus making the proposed limits more difficult to meet. The test procedure is such that the engine is operated at conditions near those which give the maximum noise level while the vehicle moves at a road speed appropriate to urban driving; thus there is considerable pressure on vehicle and engine manufacturers to reduce engine noise.

At Lucas CAV, researchers have been working intensively for 20 years to develop noise control technology for diesel engines. The techniques they have developed to control engine noise radiation have been widely reported [2–5] and a few production engines now incorporate some of this technology. During the course of recent investigation into combustion, they had cause to monitor combustion noise levels during emission tests and plot 'trade-off' curves between noise, emissions and economy [6]. The aim of the work was to find which, if any, modifications to the combustion system gave improvements in noise without significant penalties in gaseous emissions, particulates, unburnt hydrocarbons and fuel consumption, bearing in mind that all these parameters are subject to legislation in one or mor major markets.

Diesel engine noise originates as mechanical impacts and cylinder pressure pulses which are generated during the normal operation of the engine. These excite the external surfaces of the engine into resonant vibration as summarized in Fig. 1. The several sources each distort the engine structure in a different way, but all cause the external surface to deflect, vibrate in mechanical resonance and radiate noise, as a result of this motion.

2. DESCRIPTION OF NOISE GENERATION BY THE COMBUSTION PROCESS

Noise which is radiated by vibration of the engine as a direct result of the rapidly changing pressures in the combustion chambers of the cylinders is known as 'combustion noise'. Combustion in a compression-ignition engine is remarkably complex. Ignition does not occur at the instant when fuel is injected into the cylinder; there is a delay before the right conditions for spontaneous ignition are attained. This delay is known as the 'ignition delay period'. The ignition delay period depends upon the temperature and pressure of the air charge into which the fuel is injected, engine speed, fuel cetane number, and the way in which the fuel is distributed in the combustion chamber. During the ignition delay period, a number of complex processes take place:

1. Fuel is injected, either as discrete jets (direct injection) or as a hollow cone (indirect injection) into the air charge.
2. Depending upon the temperature, the injected fuel either breaks up into filaments and then droplets of fuel before evaporating, or evaporates directly from the liquid core of the jet. Air is entrained into the jet in both cases. Further mixing between fuel and air is achieved by the momentum imparted to the fuel, the bulk rotation (swirl) and squish imparted to the air charge during induction, and local turbulence, in proportions which are determined by the combustion system design. In some combustion systems, injected fuel reaches the combustion chamber wall where it may be evaporated by heat from the wall after which it is mixed by air motion. (An extreme

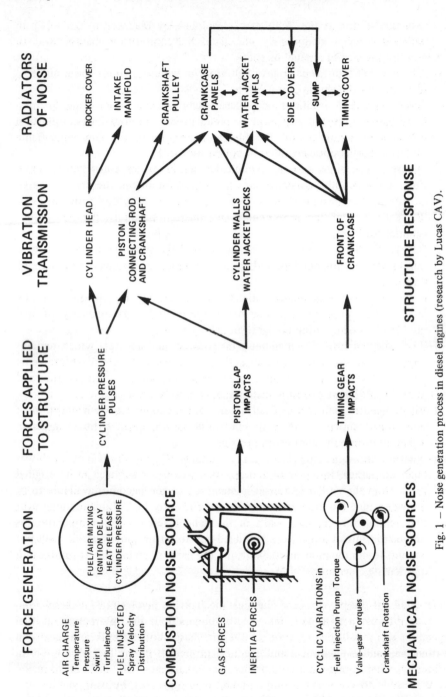

Fig. 1 – Noise generation process in diesel engines (research by Lucas CAV).

example of this is the 'M-System' developed by Meurer for MAN [7] in which the fuel is deliberately placed on the combustion chamber wall to control the rate of air/fuel mixing.)

3. The injected fuel forms zones of mixture in which air, fuel vapour and fuel droplets exist in varying proportions.

4. Several chemical reactions take place, each at a rate which depends upon local temperature and local mixture, which produce intermediate combustion products by oxidation of the fuel and give rise to the faint blue flame observed during optical studies of combustion [8].

5. Above a critical temperature which depends upon the constituents of the fuel, and when the concentration of the fuel vapour in the hot air charge is sufficient to produce a near-stoichiometric mixture, and probably with the assistance of the intermediate combustion products, the fuel ignites spontaneously.

6. The fuel which has been injected and intimately mixed with the air charge during the ignition delay period ignites and burns, giving a very rapid rate of heat release.

7. Cylinder pressure increases almost instantaneously at ignition from the compression pressure to a pressure determined by the quantity of fuel intimately mixed during the ignition delay period.

8. Fuel injected into the chamber after ignition, and any fuel which remains unburnt from the ignition delay period, is probably pyrolized before being burnt unless it is injected into an oxygen-rich zone at a temperature and pressure sufficient to ensure instantaneous vaporization.

9. All fuel burnt before top dead centre, and in some cases a proportion of that burnt shortly after, will contribute to the peak cylinder pressure achieved during the combustion process.

10. Fuel which is injected after this, the products of pyrolysis of fuel which was injected earlier, free carbon and any fuel which has survived in its original form from the earlier processes is burnt as oxygen becomes available to it, provided that the local temperature is sufficiently high. Local cooling near the combustion chamber walls or the general reduction in temperature as the piston moves away from top dead centre on the power stroke will put an end to combustion, particularly of the carbon, giving rise to smoke and particulate emissions.

The details of the physical and chemical mechanisms involved in the diesel combustion process are, as yet, imperfectly understood; the interaction of fuel injection spray ballistics, motion of the air charge both in bulk and locally in turbulence, and the physical and chemical properties of the fuel all interact in condition where precise and comprehensive measurements are difficult to make. It is possible to construct a simple model of the process, by using one or two empirically-determined factors [9] which is sufficient to examine many of the options to control combustion noise at source. Andree and Pachernegg have evolved a different theoretical model which accounts for many of the observed effects in European direct injection engines [10]. Lyn and Valdmanis have shown that compression temperature and injection timing are the major factors affecting ignition delay under normal running conditions; air velocity, injection pressure and nozzle configuration have secondary effects and injection quantities have a negligible effect [11]. The comprehensive analysis of the available engine and research data upon ignition delay, undertaken by these authors, provides both an illuminating insight into the physical factors which affect ignition delay and useful guidelines such as:

(a) Ignition delay is shortened by about 3° CA per 1000 rev/min increase in engine speed through the increase in compression temperature and pressure as well as injection pressure with engine speed.

(b) The increase in delay with reduction in load is only about 1 to 2° CA from full load to no load.

(c) The ignition delay could be reduced to a nearly constant 4° CA by fumigating with fuel introduced with the intake air, thus effectively eliminating the chemical part of the delay for fuels of widely different cetane number. More recent work by Lucas CAV suggests that ignition delay may be reduced by increasing the dispersion of the fuel during the early stages of injection, particularly in some indirect injection combustion systems.

Fuel which is pre-mixed with the air charge during the ignition delay period gives rise to the characteristically rapid rise in pressure at ignition. The effect of varying the rate of pressure rise at ignition upon combustion noise can be seen in Fig. 2. Computer-generated cylinder pressure diagrams, using a compression and expansion curve appropriate for a 16:1 compression ratio, were fed into an analysis program for real cylinder pressure diagrams recorded from actual engines. The analysis program transformed each cylinder pressure cycle into a series of harmonics in the frequency domain, and multiplied each harmonic by the appropriate component of the modulus of the transfer function of the engine and test cell acoustics, to give the noise spectrum due to combustion (1 m from the engine as measured in a test cell). This spectrum was A-weighted and the components added on a power basis to give the combustion contribution to noise 1 m from the engine in a test cell. The rate of pressure rise was varied from 3.5 MN/m^2 per millisecond to 16 MN/m^2 with all the pressure rises passing through the same mid-points. The sharp corners were deliberately smoothed, although a period of dwell was introduced at peak cylinder pressure for the higher rates of pressure rise, as shown by 'bb' in the inset diagram in Fig. 2. This procedure was repeated for two peak cylinder pressures, 10 MN/m^2 (1450 lb/sq. in.) and 9 MN/m^2 (1305 lb/sq. in.). Over much of the range of rates of pressure rise, peak cylinder pressure has little effect and combustion noise is nearly proportional to rate of pressure rise. However, at bery high rates and

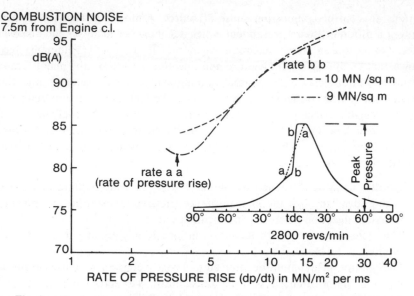

Fig. 2 – Combustion noise from cylinder pressure diagrams with various rates of
pressure rise and the same peak pressure.

even more so at low rates of pressure rise, simulating quiet combustion systems
(at and near rate 'aa' in the diagram) peak cylinder pressure controls the combustion noise level. This analysis shows the very wide range in combustion noise
level which may be obtained simply by changing the slope of the rise in pressure
upon ignition; *it shows also how all aspects of the cylinder pressure diagram
must be taken into account when quantifying the combustion noise from 'quiet'
combustion systems.*

The analysis of theoretical cylinder pressure diagrams may be taken a stage
further by generating successively smoother cylinder pressure diagrams until, for
a given engine design, a diagram is found which gives a Minimum Combustion
Noise level (MCN). MCN provides a measure of the noise reduction potential in
the combustion process of any particular engine, and it provides a reference level
by which competing quiet combustion systems may be judged. MCN depends
upon fundamental engine design parameters, and in particular compression ratio
and peak cylinder pressure, which also control the thermal efficiency.

The cylinder pressure diagram for the MCN is shown in Fig. 3(a), with that
for a noisy, naturally aspirated direct injection engine (injection starts at 20°
b.t.d.c dynamic timing) at 2000 rev/min and a lightly turbocharged direct
injection engine at the same speed. These diagrams have been analysed to give
the cylinder pressure spectra of Fig. 3(b), where the increases in low frequency
components due to the higher peak cylinder pressure of the turbocharged engine
can be seen clearly. At frequencies above 1 kHz, turbocharging goes some way to
reducing the vibration-exciting forces from the cylinder pressure development,

but it does not achieve the very low levels, at high frequencies, of the MCN diagram. Although there may appear to be a lot of potential for further reduction in combustion noise above 1 kHz, the structure response below 1 kHz of conventional engines is usually large enough for the components below 1 kHz to control the overall noise in dB(A), so the full potential of MCN cannot be exploited.

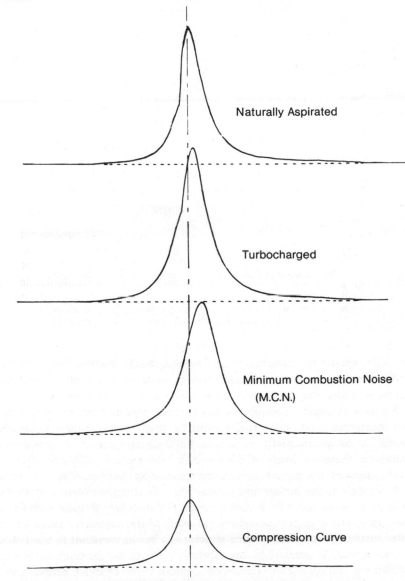

Naturally Aspirated

Turbocharged

Minimum Combustion Noise
(M.C.N.)

Compression Curve

Fig. 3(a) — Cylinder pressure developments of direct injection engines.

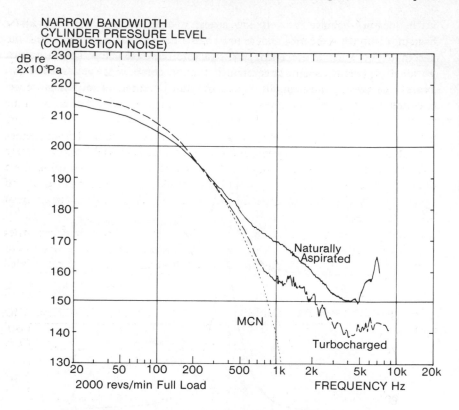

Fig. 3(b) — Constant bandwidth analysis of cylinder pressure developments in direct injection diesel engines with control of noise at source.

The engine structure has to be designed to take the combustion and inertia forces applied to the piston and moving parts of the engine. The structure deflects under the combined action of these loads as they occur, and then vibrates with decaying amplitude in a large number of normal modes. Vibration of the surface areas of the engine, which is responsible for these areas radiating noise, is composed partly of forced vibration and partly of damped natural vibration therefore. Many of the normal modes include relatively large flexural deflections of the surface areas themselves, which both increase the amplitude of vibration at the surface and increase the efficiency with which the vibration energy is converted into acoustic energy (by matching flexural wavelengths in the panel at the natural frequencies of many of the normal modes of vibration). The total response of the engine structure to the combustion pressure development is usually assessed as the 'structure attenuation', which is the decibel difference between each frequency comonent of cylinder pressure and the component of the noise 1 metre from the engine (at that same frequency). This is the decibel equivalent of the reciprocal of the modulus of the Transfer Function between cylinder pressure and noise 1 metre from the engine. With the advent of digital signal processing instruments with a dynamic range in excess of 90 dB, it is perhaps more convenient to work in terms of the modulus of this Transfer Function rather than its reciprocal, which has been termed the 'structure response' function (in decibels) although it does include a contribution from the acoustics of the test cell. Structure response curves for several engines are plotted in Fig. 4 as 1/3 octave band spectra. These were measured from engines running in conventional test cells, and the test cell acoustics add approximately 3 dB to the free field engine noise; so the structure response curves in Fig. 4 are higher by this amount than they would have been if the engine were measured at an open-air site with no acoustically reflecting surfaces present apart from the ground-plane.

The mean of these structure responses has been used by Lucas Industries Noise Centre as a standard by which to judge the noise attenuating properties of engine structures. The smoothed, mean structure response function (designated

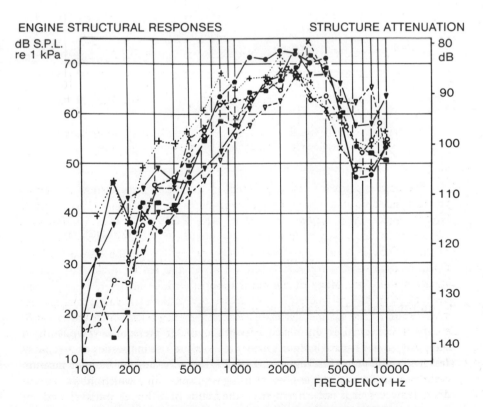

Fig. 4 — Structure responses of in-line truck and tractor engines.

SAI) is plotted in Fig. 5. This structure response function is used also to compare the noise-exciting propensity of different combustion systems and to examine the effect of degradations in fuel ignition quality upon combustion-orientated noise. Clearly a common structure response function is required if combustion systems of different engines are to be compared, and trends become more obvious if this structure response friction is a smooth continuous function of frequency.

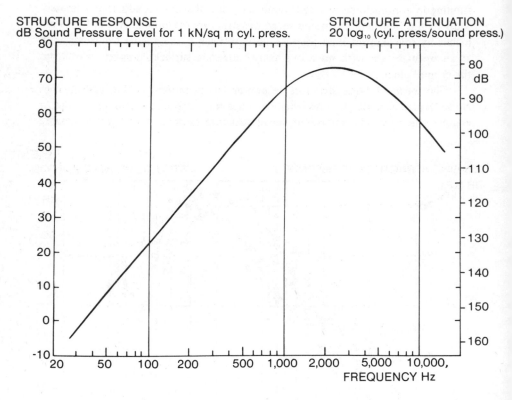

Fig. 5 – Average structure response function SA1.

3. MEASUREMENT OF COMBUSTION NOISE

The most conspicuous difference between cylinder pressure developments in diesel and ('quiet') spark ignition engines is the rapid rise in pressure immediately after ignition in most diesel engines, so that it seems natural to use some measure of this rise to determine the noise-exciting propensity of diesel combustion processes. However, it is not evident from the diagrams of Fig. 3(a) that the rate of pressure rise is far from constant during the initial rapid rise, nor is it constant from cycle to cycle in the same cylinder. Graphical methods of measuring rate of pressure rise using pictures taken from an oscilloscope or an $X-Y$ recorder are just not accurate enough to measure the characteristics of the cylinder pressure development which control the noise originating from the combustion process. The accuracy of oscilloscopes and $X-Y$ recorders rarely approaches 1%, whereas the features of the cylinder pressure development which control combustion noise may be less than 1/1000th of the total diagram amplitude! For similar reasons, instruments which employ digital sampling so that the waveform can be stored in a digital store must have an unusual combination of high resolution and rapid sampling (analogue-to-digital converters must have a voltage resolution and linearity better than one part in 2000, which requires a 12-bit ADC with better than 0.05% linearity for combustion noise work, they must also be capable of taking a new sample each 20 microseconds, to be capable of measuring events which occur once per degree at 3000 rev/min).

A digital sampling system capable of acquiring and storing data with such precision and at such a rate is expensive. An alternative is an analogue electronic differentiation circuit (5 dB/octave high-pass filter) which maintains 0.05% linearity over 4 decades with less than 1° phase change between 100 and 10 000 Hz. This is not an easy circuit to design and construct. (However, similar analogue circuits have been used to make the best use of the available signal-to-noise ratio of instrumentation tape recorders, employing frequency modulated recording techniques, for some years.) In summary, it is not easy to measure the rapid rate of pressure rise which follows ignition in most diesel engines. In addition, as Fig. 2 shows, the peak cylinder pressure has a significant influence upon combustion noise, and controls combustion noise from smooth pressure developments such as those achieved by indirect injection and turbocharged engines at high speed, full load operating conditions.

Historically, the first real advance towards precisely measuring noise originating from the combustion process was made by Priede and Austen in the middle 1950s [2]. At that time, CAV had succeeded in making a linear cylinder pressure gauge robust enough to stand up to the harsh environment of the combustion chamber, so that its diaphragm could be mounted flush with the combustion chamber wall, together with an amplifier with sufficient signal-to-noise ratio to allow the high frequency harmonics of the cylinder pressure development to be measured by a Muirhead Pametrada wave analyser with a 1.2% bandwidth. An example of one of these analyses is shown in Fig. 6, taken from [12].

Austen and Priede used-the spectra of cylinder pressure to measure the noise-exciting propensity of different combustion systems, to examine the effects of injection timing, load and speed and to examine some simple changes to the fuel injection system in a classic series of papers [2, 12–14]. One of the most significant results from this work is the effect of injection timing upon combustion noise and the definition of a 'critical cylinder pressure level curve' such that mechanical noise predominated at timings more retarded than this

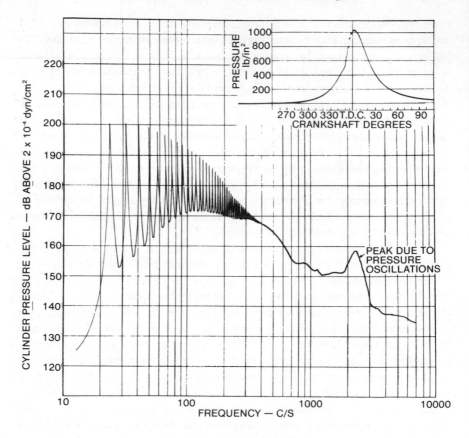

Fig. 6 – Cylinder pressure oscillogram and spectrum.

spectrum whereas combustion noise predominated at timings more advanced than this spectrum. The original Austen and Priede diagram is reproduced as Fig. 7.

   Although the spectra measured with the Muirhead Pametrada analyser allowed measurements to be made of critical parts of the combustion pressure development, each cylinder pressure analysis took 30 minutes, and comparable engine noise spectra took over 1 hour each (which is one reason why octave band spectra of noise are compared with 1.2% bandwidth spectra of cylinder pressure in most of the Austen and Priede papers). Furthermore, the injection timing of the 'critical cylinder pressure level curve' can vary across the frequency range of interest.

   Several methods have been used to measure the noise exciting propensity of diesel cylinder pressure developments, starting from the amplified signal from a cylinder pressure gauge in each case:

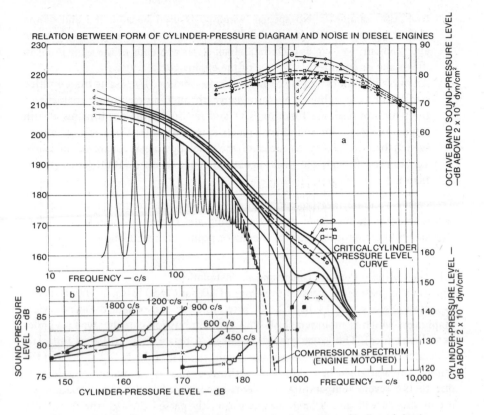

RELATION BETWEEN FORM OF CYLINDER-PRESSURE DIAGRAM AND NOISE IN DIESEL ENGINES

Fig. 7 – Relation between cylinder pressure and noise spectra of indirect injection
engine A at 2000 rev/min full load.

(a) 1.2% (constant percentage bandwidth) analysis of amplified signal from
     cylinder pressure transducer with a Muirhead Pametrada wave analyser.

(b) 1/3 octave (constant percentage bandwidth) analysis of cylinder pressure
     signal with two Bruel and Kjaer 1/3 octave band filters in tandem and a
     type 2606 microphone amplifier.

(c) Constant bandwidth analysis of cylinder pressure signal with Nicolet type
     440A 'real time analyser' with data samples which were not synchronized
     with the engine cycle.

(d) Constant bandwidth analysis with a Fourier Analysis program in a digital
     computer (D.E.C. PDP 11/60). Data blocks are each exactly one engine
     cycle long (four strokes). Measurements are the average of 20 to 100 data
     blocks, each of which comprises 1024 or 2048 digital samples of the
     cylinder pressure waveform. The analogue-to-digital converter provides
     12-bit samples (resolution 1 part in 4096) at precisely controlled rates up

to 52 000 samples/s. No special 'window' is used because the end of the cylinder pressure data block should exactly match the beginning in value and gradient, so long as each data block starts and ends at the same point in the intake or compression stroke of the engine cycle.

(e) An overall 'Combustion Noise Level' can be obtained by passing the amplified signal from the cylinder pressure gauge through an electronic analogue of the structure response function SA1 and measuring the resulting pulse with a type 193 precision sound level meter from Computer Engineering Limited (which has the ability to measure the true root mean square of pulse-like signals even though the peak signal excursion is up to 1000 times greater than the r.m.s. value).

To achieve accurate measurements of combustion noise from a cylinder pressure development requires much care in choosing and operating some sophisticated experimental techniques. Some of the major points are listed below.

3.1 The Cylinder Pressure Gauge and its Installation

The gauge chosen must measure pressures of 10 bar (1450 lb/sq. in.) for naturally aspirated engines and 14 bar (2030 lb/sq. in.) for turbocharged engines, via a diaphragm which is mounted flush with the combustion chamber wall (cylinder head lower face). The gauge should be linear to better than ±0.1% of full scale deflection, and any resonances should be above 50 kHz. It should withstand the temperatures attained in combustion without change in calibration or zero-shift due to the severe temperature transients experienced in diesel combustion. (A lot of gauges will give a large output if quickly passed through the flame of a bunsen burner which 'flame-sensitivity' seriously detracts from their accuracy.) No gauge completely fulfils these requirements yet. One of the gauges which most closely approaches these requirements has a 'heat-shield' mounted in front of the diaphragm (Kistler 6121), which reduces the output from thermal transients but the annular space between the diaphragm and the heat shield fills up with hard carbon after a few hours use, so the gauge has to be removed and thoroughly cleaned after a few hours use (every 2 hours in the sootiest engine yet encountered). It is tempting to mount the gauge at the end of a short passage to shield it from temperature transients, but acoustic resonances in the passage (Helmholz and quarter-wave-in-pipe) will add large oscillatory components to the pressure applied to the diaphragm (in addition to the acoustic resonances of the combustion chamber itself). It is current practice to mount the gauge with its diaphragm flush with the lower surface of the cylinder head, within the perimeter of the combustion bowl, even if this means that the gauge must be at a slight angle (not quite perpendicular to the cylinder head lower face). The Kistler 6121 gauge seems to suffer changes due to differential thermal expansion if clamped too rigidly, and some resilience seems to be desirable in the clamp. Gauge-to-lead connections have an irritating habit of working loose, so it is best to seal these with silicone rubber (cold curing 'silastomer'). Current types of piezo-electric gauge employ elements with a very low capacitance, and so the charge generated by the element is correspondingly small; and it is essential that all electrical connection are kept scrupulously clean therefore (a pure Freon aerosol spray seems efficacious).

3.2 The Cable Connecting the Cylinder Pressure Gauge to its Preamplifier
The cable connecting a piezo-electric cylinder pressure guage to its conditioning preamplifier has a capacitance and resistance which shunt and leak the signal from the gauge, and most coaxial cables generate a small but significant charge them-selves when they are bent or shaken. To minimize these effects, the manufacturer of the gauge supplies special low capacitance, low noise cables and connectors which must be used exclusively between the gauge and its preamplifier.

3.3 Preamplifiers for the Cylinder Pressure Gauge
Electronic preamplifiers for piezo-electric cylinder pressure gauges are designed to detect and amplify the tiny charge developed in the piezo-electric gauge element with a circuit which has an extremely high input impedance ($>1\,000\,000\,000\,000$ ohms). The input stage of modern preamplifiers incorporates an integrating circuit to detect charge directly. Therefore the charge amplifier supplied by the gauge manufacturer, or a high-performance charge amplifier should be used. The output impedance of the charge amplifier must be low (>200) to avoid significant changes in preamplifier output voltage when filters, tape recorders, oscilloscopes, etc. are connected to this output. (This rules out some proprietary charge amplifiers.)

3.4 Combustion Noise Analyses Carried Out by Lucas Industries Noise Centre
It has become common practice at Lucas Industries Noise Centre to record cylinder pressure and its first derivative with respect to time onto magnetic tape using the frequency modulation technique. Very few recorders have the very low 'flutter', in the tape transport, needed to meet the low noise requirement ('tape noise' is irrelevant when using frequency modulation). For 15 years, Lucas used a specially built Ampex FR 1300 tape recorder which achieved 54 dB signal-to-noise ratio with a bandwidth of DC to 20 kHz when operating at 60 in./s tape speed, which was adequate for direct injection engine cylinder pressure record-ings. However, for quiet combustion systems, the high frequency components were extracted by an analogue filter amplified further and recorded on a second track. A typical recording layout is shown in Fig. 8. More recently, Racal Store 7 and Store 14 have replaced the FR 1300, largely because the signal-to-noise ratio is maintained over a wider range of tape speeds. This allows the tape to be replayed more slowly when data are being fed into the digital computer, allow-ing four channels of information to pass through the analogue-to-digital converter without exceeding the sampling rate of the 12-bit ADC. Data from all four

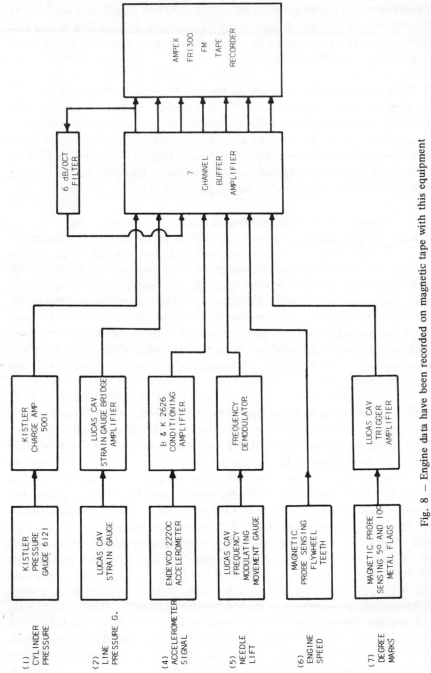

Fig. 8 – Engine data have been recorded on magnetic tape with this equipment for many years.

channels are sampled simultaneously by four 'sample-and-hold' preamplifiers; a multiplexer then directs each sample in turn to the ADC which is read continuously by the computer, which stores each new number when the ADC signals that a conversion has been made.

The data read into the computer may be stored on disk, analysed immediately or directed towards other calculation routines so that combustion noise, rate of heat release and rate of injection are all available for each captured cycle, to a common time (and crank angle) base. Each data block (1024 or 2048 samples of each channel) is triggered from the non-firing t.d.c. synchronization pulse, after a predetermined delay, if required. The sample extends over 720 degrees crank angle (sometimes 360° is chosen) so that the data are truly periodic and no window need be used ('windowing' throws away some of the data and reduces resolution).

4. SIMPLE COMBUSTION NOISE METER

To enable the noise-exciting propensity of different combustion systems to be compared, it was desirable that a simple meter, displaying combustion noise in dB(A), be connected, directly if possible, to a pressure gauge with its diaphragm flush with the wall of the combustion chamber. The signal requires some further amplification before being passed through an electronic filter which simulates the mean engine structure response (SA1) to optimize the signal-to-noise ratio of the signal output from the filter.

The structural response of the engine is impractical to simulate exactly by electronic means, but this is also unnecessary, provided the energy content of the filter response and engine structure response are the same at each frequency, the only difference will be in the waveform of the two signals. If a meter is used which can accurately measure the energy in the electronic signal, the differences in waveform are unimportant.

The peak signal level is much higher than the root mean square value as shown by the SA1-filetered cylinder pressure development in Fig. 9. A further requirement for the meter was that it must be able to handle high peak signal levels and register the mean square level accurately.

The frequency components of cylinder pressure cover the frequency range from 5 Hz to above the top of the audio frequency range, but since only audible noise is of interest, the frequency range from 20 Hz to 18 kHz is adequate. The meter should be capable of resolving levels to ±0.1 dB(A) although for most purposes, readings accurate to ±0.3 dB(A) are adequate.

The meter which appeared to meet most of these requirements at reasonable cost was the type 193 Integrating Sound Level Meter from CEL. Use of such a meter for combustion noise was attractive, since it could also be coupled to a microphone to measure sound level near the engine.

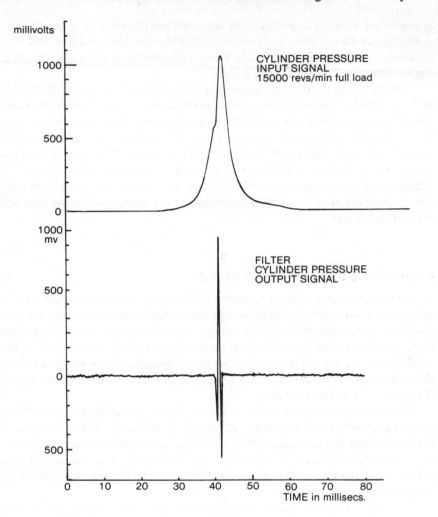

Fig. 9 — Electrical signals received and put out by structure response filter SA1.

During the past 3 years, combustion meter readings have been taken at the same engine operating conditions as those for which tape recordings of cylinder pressure have been made. The combustion noise meter results have been compared with results computed from four sets of tape recordings (from three engines) in Fig. 10. Of the 150 results, only 41 are more than 0.5 dB(A) away from the mean line and 14 were more than 1 dB(A) away (seven of these are rogue points where the CEL 193 overloaded in some way at the top of the measurement range). The combustion noise meter offers a simple technique to monitor combustion noise once a cylinder pressure gauge has been fitted. The

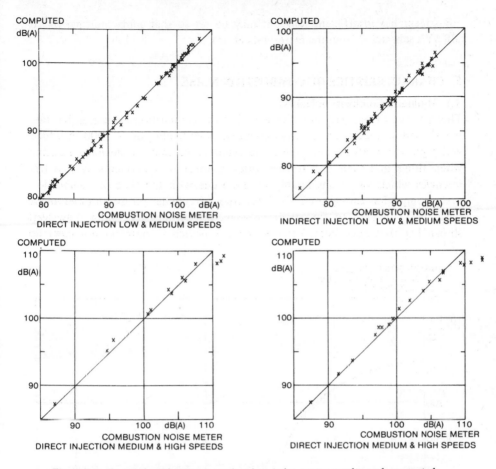

Fig. 10 – Comparison between combustion noise meter results and computed results.

direct-reading scale can be used to monitor combustion noise levels during performance and emissions testing. Although this device was developed to compare noise from different designs of combustion chamber, a calibration may be carried out and readings used to plot trade-offs of the type described in [6].

Calibration may be carried out by running the engine with advanced timing and with low cetane fuel to make the combustion noise predominant. Readings of total A-weighted engine noise may be made with the sound level meter microphone one meter from the engine surface (opposite from the mid-point of the engine) whilst combustion noise is predominant. Then with the meter connected to the cylinder pressure gauge, and with the combustion noise filter in circuit, the gain adjustment of the meter may be set up to give the same readings when cylinder pressure signals are fed to the meter input circuit (if the gain adjustment has insufficient range, it may be set so that some multiple of 10 dB(A) is subtracted from the meter indication).

5. CHARACTERISTICS OF COMBUSTION NOISE

5.1 Minimum Combustion Noise

The spectra in Figs. 3(b) and 7 suggest that the combustion excitation has two parts: there is a series of spectrum components extending up to 1000 Hz which varies by only 9 dB between the compression curve and a turbocharged engine when firing; and there is a further series of spectral components at high frequencies which vary considerably in amplitude with the type of combustion system, injection timing, etc. Various shapes of cylinder pressure development have been analysed to find their noise-exciting propensity, and it has been shown [5] that quiet combustion systems approach a compression curve, the

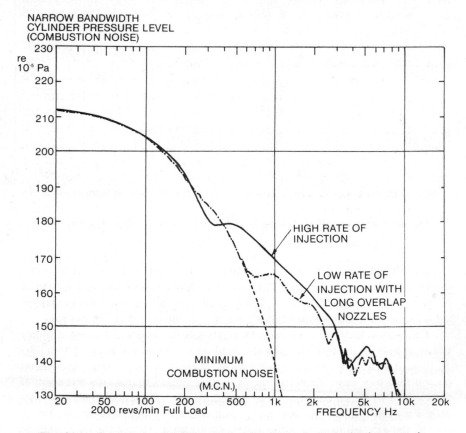

Fig. 11 – Constant bandwidth anlaysis of cylinder pressure developments in indirect injection diesel engines with control of noise at source.

components of which have been multiplied so that it has the same peak cylinder pressure.

Such a factored compression curve is the Minimum Combustion asymptote to which tend all quiet combustion systems (which have the same peak cylinder pressure). Some actual cylinder pressure spectra (envelopes of the spectral components) are compared with the appropriate Minimum Combustion Noise (MCN) spectrum in Fig. 11, for an indirect injection engine.

The spectrum for a low rate of injection deviates from the MCN spectrum above 600 Hz, and the spectrum for a high rate of injection deviates from MCN above 400 Hz. The deviation observed in the spectrum for 'high rate of injection' between 100 Hz and 400 Hz in Fig. 11 is due to a very retarded ignition which produces a double peak shape to the cylinder pressure diagram (compression then combustion), which gives rise to an increase in the harmonics close to the reciprocal of the peak spacing preceded by partial cancellation of neighbouring harmonics. Thus the spectrum falls below MCN over a limited frequency range with retarded ignition, which is compensated by low 'humps' on either side of the cancellation region.

The relative importance of the low and high frequency components of the cylinder pressure development can be judged from Fig. 12.

The upper spectra in Fig. 12 are the combustion noise contribution to overall noise of a direct injection engine with the average structure response (SA1). These spectra are unweighted; if an A-weighting is applied to simulate the effect of noise upon human ears, the spectral levels are reduced increasingly at frequencies below 1 kHz (-3 dB at 500 Hz, -8 dB at 250 Hz, -16 dB at 125 Hz and -26 dB at 63 Hz). Both the structure response and the internationally-standardized A-weighting emphasize the frequency components in the 800 Hz to 6 kHz frequency range, which is where 'quiet combustion' has most effect on the cylinder pressure spectrum.

Figure 12 shows the spectrum of an actual, noisy, cylinder pressure diagram from a 16:1 compression ratio engine with the fuel injection timing advanced to 20° crank angle b.t.d.c., and the cylinder pressure spectrum for a turbocharged engine with 16:1 compression ratio (labelled 'quiet combustion' in Fig. 12).

The MCN spectrum will be reduced if the peak cylinder pressure is reduced by $20 \log_{10} (P/10)$ dB, where P is the reduced peak cylinder pressure in MN/m^2, although this usually implies some reduction in the thermal efficiency of the cycle. Alternatively, it is possible to obtain a broader cylinder pressure diagram, than has been assumed for this MCN in Fig. 12, by increasing the cycle temperatures and allowing fuel to burn comparatively late in the cycle. This may be done with some engines at high speeds near full load conditions. However, in their current conventional form, both these types of engine tend to have high peak pressures, which detracts from the quieter combustion process.

The MCN spectrum and the A-weighted overall noise contribution which would be radiated by the engine surfaces, in response to the MCN pressure

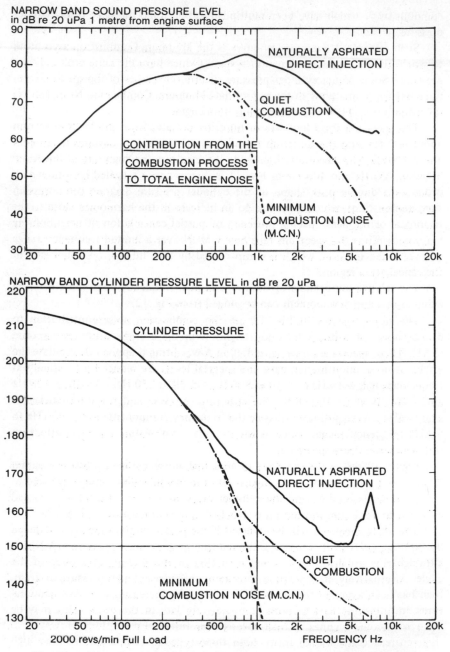

Fig. 12 – Comparison of noisy and quiet direct injection combustion processes with an average structure response (SA1). Constant bandwidth spectral analysis of cylinder pressure (lower graph) and the combustion contribution to the total noise 1 m from an engine (upper graph).

developments, provide useful reference values for actual cylinder pressure data. If noise from diesel engines is to be controlled by smoothing their cylinder pressure development, then comparison with the MCN value is essential to monitor progress and potential for further improvement.

5.2 Effect of Injection Timing on Combustion Noise

The effect of injection timing upon the high frequency components of the cylinder pressure development is shown in Fig. 13. As injection timing is retarded towards top dead centre, the ignition delay period reduces because the temperature of the air charge increases, so the fuel evaporates more quickly and the chemical production of intermediate products from the fuel takes place more rapidly. In the case of indirect injection engines, 'squish lip' or 'turbulence lip' combustion systems, the micro mixing of fuel vapour and air proceeds more rapidly as the piston is closer to top dead centre during the ignition delay. The reduction in ignition delay results in less fuel being intimately mixed with the air charge at ignition, so the peak rate of heat release and the combustion noise are also reduced. The rate of reduction of combustion noise with injection timing varies from combustion system to combustion system; for example, it varies with compression ratio, rate of heat transfer to the cylinder head and piston, rate of injection spray characteristics, etc. It is not particularly fruitful to try to lay down 'rules' for the variation of combustion noise with injection timing; it is simpler, and more accurate, to measure the data for each combustion system with the combustion noise meter described earlier in this section. It is usual to measure combustion noise as an overall dB(A) value at four or five timings spanning a range of 20 crankshaft degrees for a full mapping. Typical results are shown in Fig. 13.

5.3 Effect of Load on Combustion Noise

Since combustion noise depends to a large extent on the fuel which is premixed with the air charge during the ignition delay period, which does not change much with load at most speeds, combustion noise does not change much with load when the fuel is injected at the same dynamic timing. However, load-advance is included in many fuel injection systems to reduce full load smoke and such systems give an increase in noise with load.

On over-run, vehicle fuel injection systems usually cut off all fuel to the injectior, so when the engine is accelerated after slowing down there is a very noticeable increase in noise. In engines with a long ignition delay, at or near idling conditions, the whole of the fuel delivery may enter the cylinder during the ignition delay, and any increase in fuelling is accompanied by an increase in noise.

In the case of turbocharged engines, an increase in load (i.e. fuelling) is accompanied by an increase in exhaust temperature, which drives the turbine faster. The boost pressure increases therefore, raising the charge air temperature,

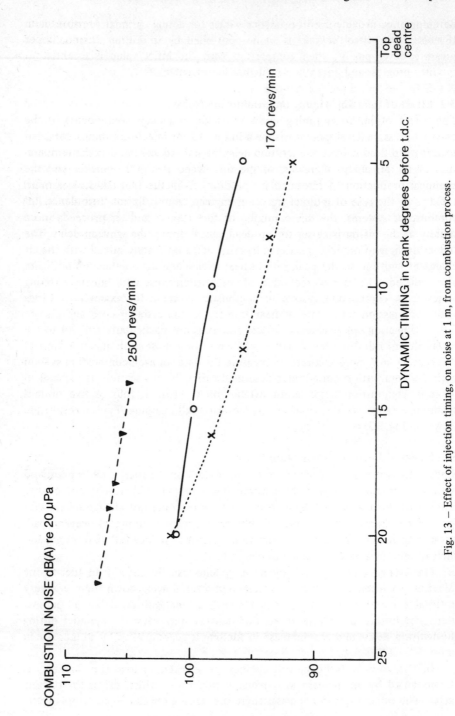

Fig. 13 — Effect of injection timing, on noise at 1 m, from combustion process.

so the ignition delay period becomes shorter. Less fuel is intimately mixed with the air at ignition so there is less combustion noise as the load on a turbocharged engine is increased.

5.4 Effect of Speed on Combustion Noise

Austen and Priede [2] plotted the increase in diesel engine noise with speed, and pointed out that the noisier the engine, the lower the rate of increase of noise with speed. If the cylinder pressure spectra are plotted together for the same engine at different speeds, not only the fundamental but all the harmonics increase in frequency in direct proportion to speed. If also the cylinder pressure development remains geometrically similar at all speeds (for the same air/fuel ratio) then the relative proportions of each harmonic will remain similar as the speed is changed. Thus as the speed is increased, the cylinder pressure spectrum will move up the frequency range with only small changes to its shape, as shown in Fig. 14.

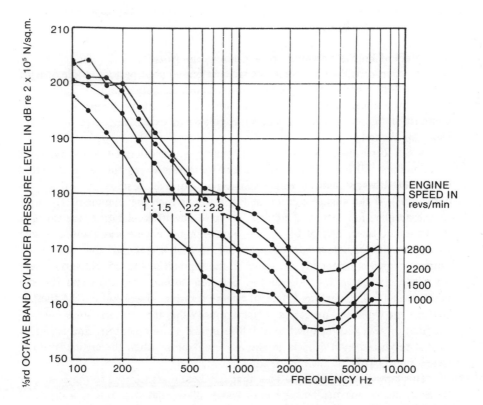

Fig. 14 — Effect of engine speed on cylinder pressure spectrum with the engine pulling full load.

The excitation to the structure at any frequency must increase with engine speeds, at the same rate as the spectrum components increase in intensity as the spectrum moves to the right. In other words, combustion noise increases with speed at the same rate as the spectrum slope. For the naturally aspirated direct injection combustion systems shown in Fig. 12, the spectrum falls at approximately 35 dB/decade from 400 Hz to 4 KHz (which encompasses most of the important structure response resonances) so the combustion increases at a rate of 35 dB/decade increase in engine speed. However, when combustion noise is reduced, the slope of spectrum may become steeper, as shown in Fig. 11, or the slope may vary with frequency, as shown in Fig. 12 ('quiet combustion').

The increase in combustion noise with engine speed is an important part of engine noise prediction formulae [15, 16].

The formulae work quite well as an approximate predictor for current conventional engines, despite assuming that mechanical noise either is negligible or increases at a similar rate as combustion noise. However, if applied to certain quiet combustion systems and for unresponsive engine structures, the increase in noise with speed will mirror the MCN cylinder pressure envelope which has a continuous change in slope.

5.5 Effect of Engine Acceleration on Combustion Noise

During urban driving, vehicle engines are accelerated after periods at idle or light load and low speed operating conditions. Such light duty operation causes the engine to become relatively cool. The influence of engine temperature upon combustion noise of direct injection engines was extensively investigated by Head and Wake [17].

A 5.8 litre direct injection truck engine was accelerated at several rates while pulling full load. The rates of speed increase covered the range expected during the drive by tests by which vehicles are normally tested (ISO 362 and its derivatives). If the engine was run at full load and high speed immediately prior to the acceleration test, the increase in noise due to acceleration was small as shown in the lower half of Fig. 15. If, however, the engine was allowed to idle for a few minutes before being accelerated, so that it was cool, the combustion noise increased considerably as shown in the top half of Fig. 15. However, there was little change in the relationship between combustion noise and ignition delay, which ruled out any effect of engine transient operation on the fuel/air mixing and ignition processes. Furthermore, the increase in noise due to acceleration was accompanied by a fall in intake air temperature, and is similar to that produced by a drop in intake air temperature when the engine is run at steady speeds with full fuelling, as shown in Fig. 16.

This strongly supports the idea that less heat is transferred to the intake air because the engine has become cool while idling, and that this is a cuase of increased combustion noise while accelerating under simulated drive-by test (and urban) engine operating conditions. In addition, less heat is transferred from the cylinder walls and piston to the air charge during the intake stroke and early parts of the compression stroke, so the charge air temperature as the fuel is injected is lower during urban driving than during high speed operation with high fuelling (e.g. motorway cruise conditions). This effect may well account for some of the differences which have been observed between engine noise measured during vehicle drive-by tests and engine noise measured on a test bed (after appropriate corrections for the different acoustic fields).

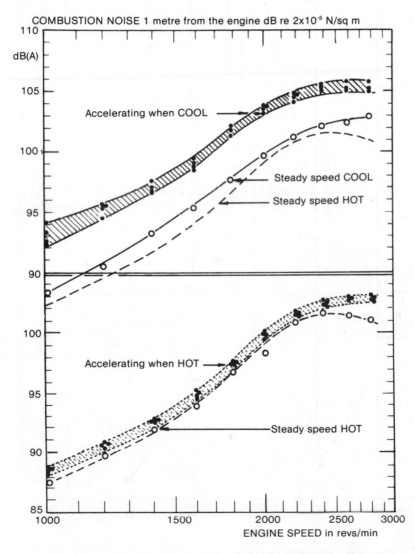

Fig. 15 − Noise of six-cylinder naturally aspirated 5.8 litre engine accelerating.

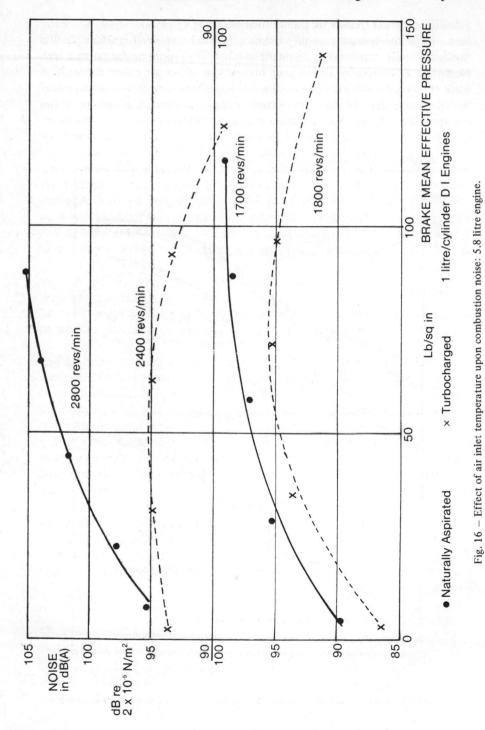

Fig. 16 – Effect of air inlet temperature upon combustion noise: 5.8 litre engine.

5.6 Effect of Fuel Quality on Combustion Noise

One of the most important properties of a diesel fuel is its ignition quality, which is usually expressed as its cetane number. The procedure for rating a fuel, by mixing reference fuels in various proportions to match either the ignition delay or the compression ratio at which auto ignition occurs, is reminiscent of the procedures for obtaining an octane number. However, the purpose of the test is almost the opposite of that to find octane numbers, in that compression ignition engine fuels are required to ignite without assistance from sparks or other energy sources. For many years, U.K. and many European diesel fuels were supplied with cetane numbers around 53 CN. However, with an increasing demand for diesel fuel and a difficult supply situation, oil companies are beginning to supply fuels with 48 to 50 CN rating instead. In North America, where diesels have to operate in very cold conditions, lighter fractions have been supplied in winter to minimize wax formation in filters and fuel lines (which prevents the flow of fuel to the engine); these lighter fractions have cetane numbers as low as 39 CN.

Low cetane fuels have longer ignition delays so they can give rise to more combustion noise. They are more prone to misfire, and in an attempt to compensate for this, operators may advance the injection timing which will extend the ignition delay period further still, and give rise to a further increase in combustion noise.

Some idea of the effect of cetane numbers may be obtained from the trade-off curves for a 1 litre/cylinder truck engine which was tested at several dynamic injection timings on 60.4 CN, 53.3 CN and 36.3 CN fuels. The trade-off curves of peak cylinder pressure, combustion noise and nitric oxide exhaust emissions all plotted against fuel consumption over a 20° crank angle range are shown in Fig. 17. (20° b.t.d.c. in steps of 5° to 0° t.d.c., advanced timings are to the top left of all curves.) The peak cylinder pressure does not change a great deal with fuel quality, because it depends mainly upon fuel injected before top dead centre. The increase in ignition delay with 36.3 CN fuel causes 4.5 dB(A) increase in combustion noise. The 16:1 c.r. engine would not run on the low cetane fuel at the more retarded timings (less than 7° b.t.d.c.) so the injection timing cannot be retarded to reduce the extra noise caused by this fuel. The high cetane fuel was obtained by adding isopropyl nitrate, which may have increased the nitric oxide emission; but there is a clear indication of worse emissions with low cetane number in the lowest graph.

5.7 Effect of Fuel Injection Rate on Combustion Noise

If the fuel injection equipment is made to inject, and mix large quantities of fuel in the ignition delay period, then an increase in combustion noise may be expected. However, if by so doing fuel consumption and the smoke are reduced to such an extent that the injection timing can be retarded without unduly increasing the smoke or the consumption, then it is possible to end up with a

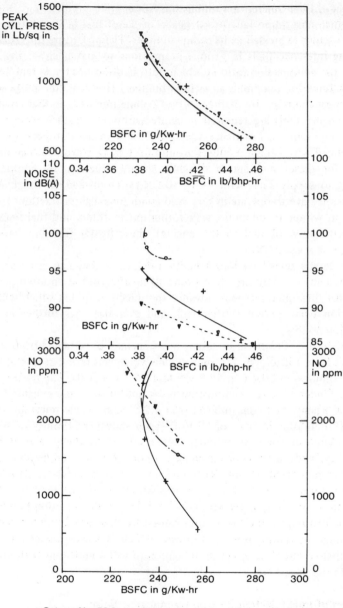

+ Cetane No. 53.3   o Cetane No. 36.3   ◁ Cetane No. 60.4

1 Litre/cyl Nat. Asp. D.I. engine DPA pump 1700 revs/min Full Load

Fig. 17 — Effect of cetane number.

better compromise on moise-v-smoke or even noise-v-consumption. Figure 18 shows trade-off curves for such an engine (naturally aspirated, directed port, 1 litre/cylinder truck, direct injection) with conventional and high rate injection equipment. Improvements like this are only possible in certain engines. The principle behind the trade-off curves is illustrated in Fig. 19. The effect of dynamic timing on combustion noise and smoke is illustrated in the left-hand graph; combustion noise is plotted against smoke separately for variations in

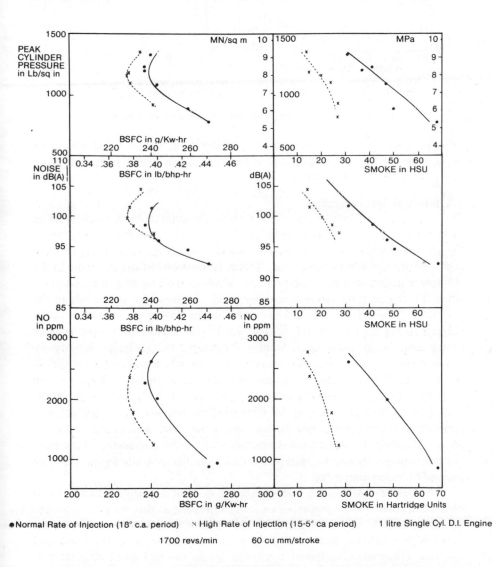

● Normal Rate of Injection (18° c.a. period)   ⅹ High Rate of Injection (15-5° ca period)   1 litre Single Cyl. D.I. Engine

1700 revs/min   60 cu mm/stroke

Fig. 18 – Effect of increasing rate of injection into a naturally aspirated engine.

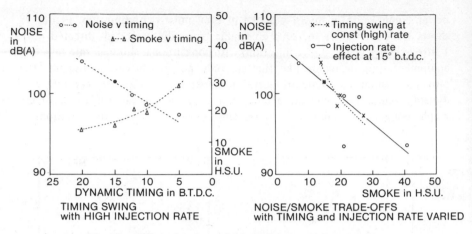

Fig. 19 — Effect of timing and injection rate on noise-v-smoke.

dynamic injection timing and mean rate of injection. The fact that the timing swing trade-off has the steeper slope is the basis for the overall improvement by increasing mean injection rate.

5.8 Effect of Turbocharging

The main effect of introducing more air into the cylinder by supercharging or turbocharging an engine (without intercooling) is to increase the temperature of the air charge when the fuel is injected, so reducing the ignition delay, premixed fuel quantity and combustion noise. This only works when the air intake to the cylinder is maintained at a high pressure, which in the case of a turbocharger is when the engine is operating at sustained high fuellings (and often only in the top half of the speed range). Thus under urban driving conditions, a turbocharged engine may not be appreciably quieter than a naturally aspirated engine of the same compression ratio. Some turbocharged engines have a lower compression ratio than the naturally aspirated versions, which makes them noisier in urban traffic. When a turbocharged engine is accelerated there is a delay between the engine speed rising and the intake air pressure rising while the turbocharger accelerates. Although this may be minimized by low inertia of the turbine—compressor shaft, by tailoring turbine and compressor efficiencies, by waste gates, etc., there are still transient situations where the combustion noise from the turbocharged engine is greater than that from the naturally aspirated engine becuase of 'turbocharger lag'.

The best way of using a turbocharger to quieten an engine seems to be to obtain the same or more power with a lower top speed so that the gear ratios can be (numerically) reduced. Figure 20 shows how considerable improvements between turbocharged and naturally aspirated engines at rated speed and load can fade into barely significant improvements in the mid-speed range at part load.

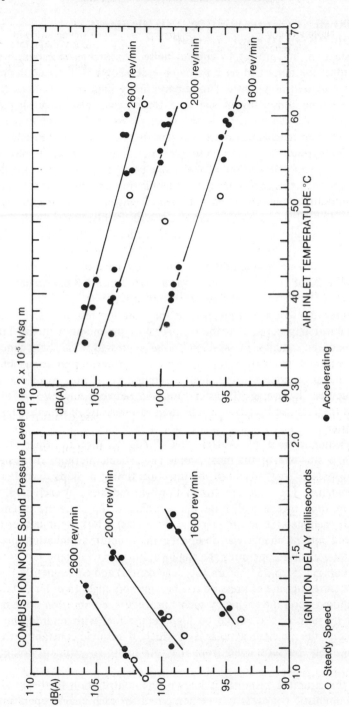

Fig. 20 — Effect of load on noise 7° CA DPA load advance.

6. MEASUREMENT OF COMBUSTION AND MECHANICAL CONTRIBUTIONS TO ENGINE NOISE

The first stage in any project to control noise at source is to establish which sources control the noise and by how much each should be reduced in order to achieve some desired noise level. This is particularly important in diesel engines where combustion noise, piston slap and timing gear rattle may all produce similar noise levels at some speeds and loads. A method for separating combustion noise from mechanical noise was reported by Austen and Priede [2]. In essence, this required combustion noise to be increased until it was predominant, so that the 'structure attenuation' of the engine could be measured as the decibel difference between the spectrum of the combustion pressure development and the spectrum of the noise measured 1 m from the engine. To do this, the cylinder pressure was sensed by a pressure transducer with its diaphragm flush with the underside of the cylinder head, that is flush with one wall of the combustion chamber.

The pressure signal was amplified, calibrated in units of sound pressure level (in decibels re 20 μPa), and the signal was analysed into 1/3 octave band levels, or 1.2% narrow band levels, as if it were a sound signal.

Structure attenuation is a measure of the vibration response of a structure to the combustion process. It is the reciprocal of the modulus of the Transfer Function between cylinder pressure and noise expressed in decibels. Since it is more conventional to analyse the dynamics of structures in terms of their response, and since sophisticated analysis equipment is now readily available which computes the average Transfer Function between any two signals, the modulus of the Transfer Function will be used as a measure of structure response in this chapter.

The technique used by Austen and Priede to measure their structure attenuation is still one of the most simple and reliable methods for measuring structure response. They advanced the injection timing in steps to increase the combustion noise. The structure attenuation was measured at each step. If the contribution from mechanical noise was significant, the structure attenuation increased at each step (structure response decreases). Thus the injection timing was advanced until no change was observed in the structure attenuation spectrum. This procedure involved operating the engine at full load with some very advanced injection timings (25 to 35° before top dead centre) and consequently the peak cylinder pressure and rate of pressure rise became very high. In order to limit the stresses imposed upon the engine by such a procedure, combustion noise may be suppressed by running the engine on high cetane fuel with retarded injection timings to find the mechanical noise level which, if assumed constant with injection timing, may be subtracted from the total noise to estimate combustion noise [4].

Once the structure response is known, the contribution to the total noise from the combustion process can be computed for each engine operating condition. This calculation requires each frequency component of the cylinder pressure spectrume to be multiplied by the corresponding frequency component of the structure response (or adding them, if both are expressed in decibels) to obtain the combustion noise spectrum for each engine condition. These compoents may be A-weighted and summed to give the combustion noise in dB(A). If the combustion noise spectrum is subtracted from the total engine noise spectrum (by converting both to sound intensities), an estimate may be made of the noise originating from mechanical sources at each engine operating condition.

Many of the engine structure response spectra measured in this way were similar in shape (±3 dB), so Lucas Industries Noise Centre built an electronic filter to simulate the average structure response function for comparative measurement of combustion noise. The output signal from this filter is measured with a soundlevel meter which is capable of registering the true root mean square value of a signal with a crest factor exceeding 25. The combination of a filter and sound level meter form a 'combustion noise meter' which can indicate the effects of future fuels or compare noise levels from different combustion chambers. For those engines which have a structure response function which is similar in shape to that of the electronic filter (shown as SA1 in Fig. 5), the meter can be calibrated in absolute units of sound level to read combustion noise directly. For such engines, the combustion noise meter and a conventional sound level meter may be used together to obtain estimates of combustion noise levels and mechanical noise levels at each engine speed and load condition. As an example, Fig. 21 shows results from a small indirect injection engine where combustion noise was increased by reducing the fuel cetane index from 56 to 46 and 39.5.

The injection pump timing remained constant throughout and the data contain additional results from alternative injector nozzle specifications. The combination of results at different speeds helps to position the 45° 'combustion noise line' even if the combustion noise meter scale is not calibrated in absolute units. The shape of the curve through each set of points was obtained simply by adding a range of intensities of combustion noise to the intensity of a constant mechanical noise source, and plotting the total intensity, expressed in decibels, against combustion noise on another piece of graph paper with the same scale intervals.

It proved to be easy to fit this curve to the measured points by eye. This simple analysis took less than one hour to complete. If it is known, or suspected, that the mechanical noise level changes with dynamic injection timing (or fuel cetane number), the shape of the curve which is fitted to the points can be modified by assuming a slight slope, or a step near t.d.c., in the mechanical noise versus injection timing relationship (which is plotted as mechanical noise versus combustion noise meter reading at that timing). If desired, the curve fitting may be performed in a computer, with a series of mechanical noise versus timing relationships, until the best fit is obtained.

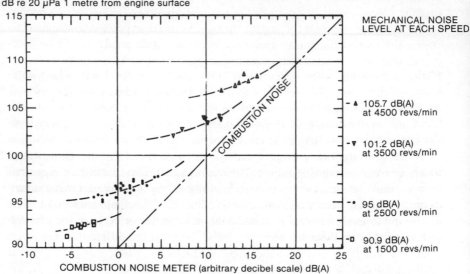

Fig. 21 – Analysis of external noise and combustion noise meter results to separate contributions from combustion and mechanical sources by changing fuel (56, 46 and 39.5 CN) and injector nozzle specification.

If the mechanical noise versus timing relationship is expressed as

$$M = mC + K \quad \text{in decibels}$$

where

M = the mechanical noise level in decibels at timing $t°$ b.t.d.c.
m = slope of the mechanical noise for increasing combustion noise (timing)
C = the combustion noise meter reading at injection timing $t°$ b.t.d.c.
K = the mechanical noise level at some chosen combustion noise level (e.g. 90 dB(A) in Fig. 2)

then the curve to be fitted to the plot of external noise measurements (E) versus combustion noise measurements (C) takes the form:
Measured Engine Noise

$$(E) = 10 \log_{10} (10^{0.1 \times M} + 10^{0.1 \times C}) \text{ decibels}$$

In intensities

$$I_E = I_M + I_C$$

and

$$I_M = K \times (I_C)^m$$

$$I_E = I_C + K(I_C)^m$$

where I_M is the sound intensity equivalent to a sound level of M dB (and similarly I_C for C dB and I_E for E dB). The best fit curves for each m, taking $m = 0$ and other small values, should be compared to find the best curve fit overall and hence the best estimate of K and the best estimate of the way in which M changes with dynamic timing (or cetane number etc.).

With the increasing availability of two-channel data-analysis systems, which provide averaged values of transfer function and coherence function between two signals, there is much interest in using sophisticated analysis techniques to separacte and rank the contributions from the several sources of noise within diesel engines. There are two major problems in the application of such techniques to diesel engines:

1. At steady speed and load conditions, with a constant injection timing and fuel quality, the combustion process, piston slap impacts, timing gear impacts, fuel injection pump torque reaction, etc. all occur at precisely consistent points in the basic engine cycle, so they are all coherent, one with another.
2. The diesel combustion process may be described as a random process which occurs periodically. It is necessary to average over a number of engine cycles to obtain statistically significant results.

The minimum of data samples is required if each data capture is triggered by a pulse generated by a flag on the flywheel of top dead centre (or b.d.c.) marking. Special triggering arrangements must be made to ensure that capture begins at the top of the exhaust stroke, and not in the middle of combustion. For noise studies, each cycle is divided into 1024 (or a multiple of 2 and 1024) equal time periods, and data from all variables are sampled simultaneously at the beginning of each time period. The analogue data are converted to 12-bit digital numbers, before being stored by the computer. Small cycle-to-cycle variations in engine speed may prevent the acquisition of contiguous data unless separate systems are used for alternate cycles. In most systems, cylinder pressure level during the intake stroke is below the zero level of the digital-to-analogue converter, owing either to AC coupling or to a deliberate offset to maximize the signal-to-noise ratio in associated instrumentation. Any form of window will produce a discontinuity in such signals and should be avoided.

Very little work has been published on the degree to which mechanical impacts from piston slap, timing gear rattle, and other mechanical sources are coherent with combustion noise in diesel engines [18]. Injection, and hence combustion, are made to occur at very precisely-determined points in the diesel cycle, and the variations observed fall within a range which is less than ½° engine rotation. Since the timing of each piston slap is determined by crank position, and a combination of gas force and inertia forces acting upon the piston, it seems very likely that piston slap and combustion noise will be coherent sources in most diesel engines. Similarly, timing gear rattles are governed by injection pump and valve gear torque variations which are precisely phased to crank rotation. It seems likely that any test for coherence between combustion and measured noise external to the engine will always show a high coherence (near unity) even if piston slap were the predominant source of noise. In order to test this proposition a piezo-electric strain gauge was attached to the water-side of a cylinder of a 1 litre/cylinder truck engine. The gauge was waterproofed so that it could provide signals proportional to cylinder deflection while the engine was operating normally. This signal and cylinder pressure from the same cylinder were recorded simultaneously via the same headstack of an FM instrumentation tape recorder (Racal Store 7, IRIG format). The engine was run at a steady speed of 1500 rev/min with 25% of full load while the recordings were made. One hundred data samples, each of 1024 digital conversions and covering exactly one engine cycle, were taken from the recorded data for analysis. The coherence function, shown at the top of Fig. 22, was nearly unity up to 2 kHz, showing that piston slap impacts are so precisely timed with respect to combustion noise at steady engine operating conditions that they appear to be the same source over much of the frequency range of interest.

In an attempt to destroy the coherence between combustion and piston slap, similar data was captured during 100 engine cycles with the engine at the same constant speed and load; but in this second experiment the injection timing was advanced progressivley, over 15° crank angle, as the tape recordings were made. In the second experiment, each data sample captured had a different injection timing to all the others. By this means, the relative-phasing of piston slap impacts to the combustion process was changed over the ensemble of data samples. The coherence function between cylinder strain (deflection) and cylinder pressure for these operating conditions is low above 200 Hz, as shown at the bottom of Fig. 22.

No attempt was made to measure the forces applied to the structure by timing gear rattle or fuel injection pump torque reaction, so it is not known whether a timing swing of 15° CA would be sufficient to separate these mechanical excitations from both combustion and piston slap, or whether they would still be coherent with one or the other. The idea of changing the relative phasing between mechanical sources and the combustion process may be extended by changing the fuel quality, intake air temperature and brake load to give a much wider variation across the ensemble of sample engine cycles captured. Once the coherence between sources has been destroyed, it begins to be possible to use multiple coherence analysis techniques to estimate the contribution from each non-coherent source. Although this approach is very attractive, meaningful

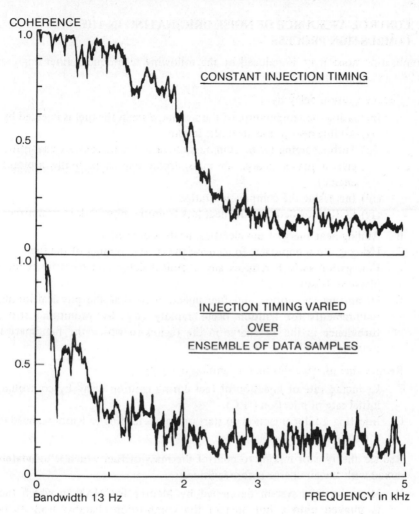

COHERENCE

CONSTANT INJECTION TIMING

INJECTION TIMING VARIED

OVER

ENSEMBLE OF DATA SAMPLES

Bandwidth 13 Hz   FREQUENCY in kHz

Fig. 22 – Reduction in coherence between combustion (cylinder pressure) and piston slap (strain in cylinder wall) caused by taking an ensemble of data samples with various injection timings.

measurements of coherence require literally thousands of data samples, where each data sample comprises at least 1024 digital conversions per engine cycle of the force applied by each potential noise source to the structure, plus a similar number of digital conversions for each vibration or noise measurement (that is at least 5120 digital numbers per data sample for between 1000 and 10 000 data samples!).

7. CONTROL AT SOURCE OF NOISE ORIGINATING IN THE COMBUSTION PROCESS

Combustion noise may be reduced by the following techniques, either singly or in certain combinations:

1. Reduce ignition delay by:

    A   Increasing the temperature of the air charge when the fuel is injected by:
        (i) Adding heat to the air in the intake.
        (ii) Turbocharging (with compression ratio unchanged, or changed to give a higher charge air mass/stroke than a naturally aspirated engine).
        (iii) Increasing the compression ratio.
        (iv) EGR, provided that the heat loss from the pipework is minimized.

    B   Adding heat energy from electrical heaters or sparks.
    C   Using ignition improvers to increase the cetane number of the fuel.
    D   Fumigation (which reduces any chemical delay and partially reduces physical delay).
    E   Organizing air motion and fuel injection so that the physical mixing requirements for ignition occur rapidly (e.g. by promoting strong turbulence in the air charge in the region to which the fuel spary is directed).

2. Reduce fuel injected during the ignition delay by:

    F   Reducing rate of injection of fuel during ignition delay by controlling initial rate of injection (IRC).
    G   Injecting a pilot injection to start processes leading to ignition ahead of the main injection.

3. Reduce the rate of mixing to control the mass of fuel which is intimately mixed with the air charge before ignition:

    H   M-combustion system developed by Meurer for MAN in which fuel is sprayed onto a hot area of the combustion chamber wall, to be evaporated by heat from the wall. The rate of mixture formation is controlled by the rate of injection, the rate of evaporation and the charge air motion (swirl).
    I   Arrange the air motion so that injected fuel mixes with a small portion of the air charge during the ignition delay period and use energy from the combustion process to ensure good air/fuel mixing during the diffusion burn phase of combustion. This is the effect of most pre-chamber and indirect injection systems.
    J   Inject fuel at a low mean spray velocity in a spray directed to a small volume of the combustion chamber during the ignition delay period followed by high rate of injection for the diffusion burn period. This might be achieved by a two-stage nozzle or by separate nozzles.

4. K   Reduce combustion rate upon ignition, for example by dilution of charge by exhaust gas or nitrogen at part load conditions [19].

5. When a smooth diagram has been achieved further reductions will depend upon the reduction of peak cylinder pressure:

   L   Time injection to achieve ignition close to t.d.c. and either:

      (i) reduce fuel injected before t.d.c., or
      (ii) reduce fuel/air mixing before t.d.c., or
      (iii) inject and mix fuel at an increasing rate shortly after t.d.c., in a direct injection engine with a high compression ratio to approach constant pressure combustion.

Many of the above are already employed in production or have been used in past designs of engine. However, items E and J above are both difficult to achieve over a wide speed and load range, and are included for completeness only.

8. ENGINE STRUCTURE RESPONSE

8.1 Response of Engine Structure to Forces

When forces are applied to engine structures by the combustion process and mechanical impacts, small, abrupt movements occur which physically displace the external surfaces periodically (forced vibration) and cause the structure to vibrate in resonance in a large number of normal modes (damped natural vibration). Many of these normal modes involve motion at right angles to the plane of the surface which forces the air in contact with them to move. At low frequencies, the air moves without significant pressure variations being generated, and little energy is imparted to the air. At higher frequencies, where the half-wavelength of sound in air is similar to the dimensions of the radiating surfaces, sufficient pressure may be built up by the vibration to cause a lot of sound power to be radiated. At higher frequencies still, the surface of the structure vibrates in a very large number of normal modes, each making a contribution to the total sound radiated by the engine. The structure response of a running engine to the cylinder pressure development was simulated by combining the response to fluctuating forces applied to the top face of the piston with the response to fluctuations in pressure applied to the underside of the cylinder head [4]. This work showed how the piston/connecting rod/crankshaft resonance could dominate the structure response to such an extent that increasing the natural frequency of this resonance moved the major peak in the structure response by the same frequency interval. In this case, the piston/connecting rod/crankshaft resonance was well coupled to the most important noise-radiating modes of the crankcase panels. It acts like a single degree of freedom system emphasizing the response of, and therefore noise radiated by, the crankcase panels and oil pan at, and near, its natural frequency. At frequencies which are 50% or more higher than this natural frequency, the response of the lower parts of the engine is much attenuated, suggesting vibration isolation.

Earlier work [3], Fig. 7 showed how motion of the cylinder head preferentially excited certain modes of the rocker cover, which then radiated noise. In the frequency range up to 6 kHz, the cylinder head underwent forced vibration with a half-wavelength of about one cylinder pitch (and not less than one cylinder diameter). Flexural modes of the rocker cover of similar wavelength were excited preferentially by this motion. It so happened that these rocker cover modes could radiate noise effectively, when the rocker cover was made from 1.2 mm (18 swg) sheet steel, 4.75 mm (3/16 in.)-thick cast-iron or 4.75 mm-thick cast aluminium.

Although published information is scarce, representatives of more than one engine manufacturer have observed differences in engine noise (piston slap) when liner design has been changed. Clearly there is scope for noise control by a careful redesign of the components of the engine responsible for transmitting vibration from each source to the external surfaces which radiate noise.

The structural components forming the external surfaces of the engine which radiate most of the engine noise are shown in Fig. 1. Many of these have numerous flexural modes of vibration, which are able to radiate noise efficiently above a certain frequency. This frequency depends upon the size of the engine, but it may be taken as approximately 600 Hz for small truck engines. The sump and valve gear covers are excited by motion of the crankcase/cylinder block casting transmitted through the bolted joints between them. Modifications to the crankcase/cylinder block casting can have a profound effect on the vibration of these covers and the oil pan in particular.

The crankcase and water jacket vibrate in a series of normal modes which have been intensively investigated [3, 4, 26]. At high frequencies, groups of closely-related mode shapes appear, with only small differences in mode shape and natural frequency between the members of each group. Common normal modes for in-line engines [4] are:

1.   Crankcase bending with torsion of cylinder block.
2.   Fundamental crankcase panel mode. With respect to any reference panel, the adjacent panels on that side of the engine, and the panel opposite on the other side of the engine, move 180° out of phase to the reference panel. This mode is easily excited by the cylinder pressure development via the main bearing diaphragms, which move axially in sympathy.
3.   Skirt-flapping modes on long-skirted crankcases.
4.   Water jacket flexural modes

V-form engines have been shown by Lalor [27] to possess additional important modes:

5.   Torsional modes of the whole crankcase and cylinder block.
6.   'Tuning fork' mode of cylinder block and cylinder heads.

Most of these modes are capable of radiating a significant contribution to the overall noise of a conventional truck engine.

The normal modes described above are fundamental to the response of the structure to mechanical impacts as well as to cylinder pressure development, so their control offers scope for reducing mechanical and combustion noise together. The distortion of the crankcase/cylinder block casting (under forced vibration and natural damped vibration of the above normal modes) excites forced and resonant vibration in the oil pan and valve gear covers. Hence, if noise is to be reduced by modifications to the surface structures of the engine, it is essential that control of crankcase/cylinder block movements be treated as a primary objective [3, 27].

8.2 Control of Structure Response

Control of the vibration of the external surfaces of the engine offers a way to control both combustion noise and mechanical noise.

10 dB(A) reductions in structure response were demonstrated by Priede *et al.* between 1959 and 1963 [12], and two crankcases with reduced structure response, which Lucas CAV helped to design, have been in production for 9 years [3].

There are three different approaches to control of structure response, which may be summarized as:

1. Reduction of the amplitude of vibration of natural damped vibration of those normal modes of surface structures responsible for radiating most noise [3–6, 28] to give the reduced structure responses shown in Fig. 23 by:
   (a) Tuning major noise-radiating modes of crankcase/cylinder block castings to frequencies where excitation is least. Plus
   (b) Applying constrained-layer vibration-damping treatments to oil pan and valve gear covers if these are made of flexible sheet metal. Plus
   (c) Applying vibration isolation treatments to oil pan, valve gear covers, crankshaft pulley and intake manifold where these are flexurally stiff (castings or fabrications).
2. The redesign of the engine structure along the lines of the Structure Research Engines constructed by Priede *et al.* [12].
3. Encapsulation of the engine in a close-fitting shell made from flexible, highly damped sheet material, and isolated from all high frequency engine vibration. Thien [29] has achieved 20 dB(A) reductions with thin steel encapsulation techniques which leave too little space between shell and engine for acoustic resonances to be set up, thus obviating the need for acoustic absorbent in this space.

Of these three approaches, the first is the most attractive for high volume production, as it requires the minimum of new processes and manufacturing plant, and it avoids the increased cost of maintaining a complete encapsulation 'airtight'

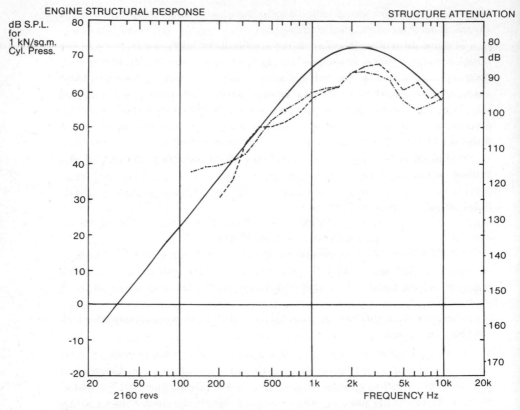

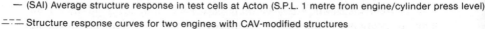

— (SAI) Average structure response in test cells at Acton (S.P.L. 1 metre from engine/cylinder press level)

--- Structure response curves for two engines with CAV-modified structures

Fig. 23 – Structure response functions for conventional and quiet structures.

in service. Following a very thorough investigation of the vibration of the crankcase, cylinder block, sump and valve gear covers, modifications were carried out on four engines by Lucas Industries Noise Centre [3].

In all four cases it was found that the major noise-radiating modes of the crankcase and water jacket (first panel modes) were efficient radiators of noise. Rib patterns were evolved for these panels to tune them to frquencies where excitation from the cylinder pressure development was least. Sheet steel oil pans and covers were modified to be as flexible as possible and a constrained layer damping treatment was bonded onto them in such a way as to maximize the energy loss in the normal modes which radiated most noise. The stiff cast aluminium and cast-iron covers were fitted with vibration isolation devices to prevent excitation of modes which could radiate noise efficiently; the crankshaft pulley was similarly treated. When all the treatments were fitted to the four engines reductions of 5 to 8 dB(A) were achieved together with a marked reduction in the harshness of the sound, which was ascribed to the removal of some of the high frequency tonal components. The structure response curves of two of these engines are compared to the average response curve of eight current production engines in Fig. 23. The average response curve offers a convenient comparison for all automotive diesel engine structures with a swept volume of between 1.5 and 10 litres, most of which have very similar structure response functions [5] Fig. 5.

The improvements in structure response of the modified engines over the average response for untreated engines are most marked in the frequency range 700 Hz to 5000 Hz, which is also where the human ear is most sensitive at low levels of sound.

9. COMBINATION OF TREATMENTS TO CONTROL NOISE AT SOURCE

If a quiet combustion system is installed in an engine with a low (quiet) structure response, considerable reduction in high frequency noise may be achieved, as shown in the upper graph of Fig. 24. However, neither technique reduces the noise much below 500 Hz, so components at these frequencies will tend to influence the overall noise level more and more as the noise is reduced. There seems to be little chance of reducing combustion excitation components at these frequencies without compromising the torque or efficiency of the engine. However, the structure response may be reduced at these frequencies by further attenuation to the oil pan and front cover (and the intake silencer and exhaust system, of course). Even so, when quieter combustion, achieved by turbocharging, is combined with the structural modifications developed to reduce structure response, large reductions in the high frequency noise will be achieved, which will make the engine sound much quieter subjectively.

The overall noise reduction in dB(A), available if a quieter combustion spectrum is combined with the best (smallest) structure response function, is shown in Fig. 25. The combustion noise of a quiet combustion system in an average structure (SA1) is also compared with the same combustion system in a quiet structure. The lowest level of mechanical noise which has been measured is included together with an assumed level in a quiet structure. Figure 25 shows that under steady speed (hot) operating conditions, the best combination just meets the noise levels which may be required of engines to go into trucks in the late 1980s.

The engine which incorporates quiet combustion processes and minimum mechanical noise in an unresponsive structure has not yet been built, so the predictions shown in Fig. 25 were obtained by combining actual and calculated cylinder pressure data with actual structure response functions in a digital computer. The computer was programmed to calculate the A-weighted combustion

NARROW BAND SOUND PRESSURE LEVEL
in dB re 20 µPa 1 metre from engine surface

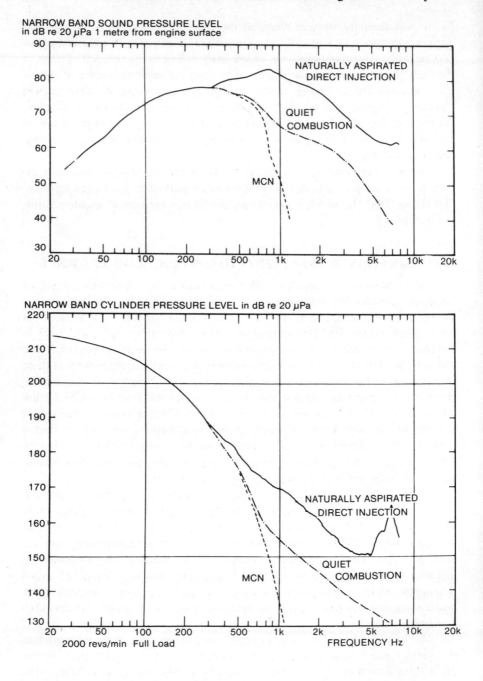

NARROW BAND CYLINDER PRESSURE LEVEL in dB re 20 µPa

2000 revs/min Full Load

FREQUENCY Hz

Fig. 24 – Combination of noisy and quiet combustion processes with an average
engine structure response.

NOISE 7.5 metres from engine centre line

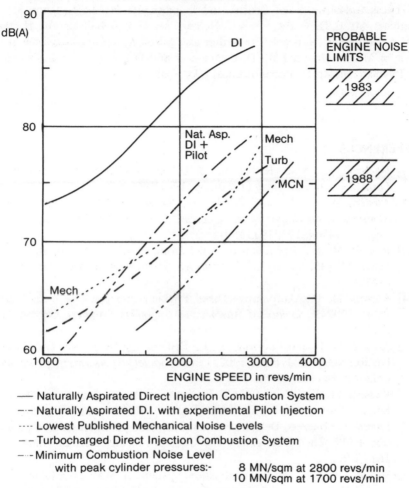

- Naturally Aspirated Direct Injection Combustion System
--- Naturally Aspirated D.I. with experimental Pilot Injection
···· Lowest Published Mechanical Noise Levels
- - Turbocharged Direct Injection Combustion System
---- Minimum Combustion Noise Level
     with peak cylinder pressures:-   8 MN/sqm at 2800 revs/min
                                          10 MN/sqm at 1700 revs/min

Fig. 25 - Simulated combustion noise levels for 1 litre/cylinder engines.

noise (via a spectrum of the combustion originated contribution to engine noise) at several speeds. With the inherent facility to change peak cylinder pressures, structure response, heat release diagram, etc., this set of computer programs represents a very advanced noise-prediction tool.

ACKNOWLEDGEMENTS

The author is grateful to the directors of Lucas Industries p.l.c. and Lucas CAV Limited for permission to publish this material.

The work contains important contributions from past and present members of Lucas Industries Noise Centre, and in particular Mr. D. J. Amos, Mr. E. Gardiner, Mr. C. D. Young, Mr. A. J. Herbert, Mr. G. P. B. Balfour, Mrs. H. Head, Mr. A. Chung, Dr. S. Nicol. The author also gratefully acknowledges the assistance of Mr. W. May and Mr. D. Lemmon of SGRD Ltd. and Mr. L. Cox, Miss S. Twyman and Mr. D. Morris of Lucas CAV Limited.

REFERENCES

[1] MacNeill, J., 'Noise Reduction—an Urgent Need', *Noise and Vibration Control World Wide*, May (1980), pp. 194—196.

[2] Austen, A. E. W., and Priede, T., 'Origins of Diesel Engine Noise', *Symposium on Engine Noise Suppression*, Oct. (1958), *Proc. Inst. Mech. Engrs., London*, **173**, 19 (1959).

[3] Russell, M. F., 'Reduction of Noise from Diesel Engine Surfaces', SAE Paper 720135, *Society of Automotive Engineers Congress*, Detroit, Jan. (1972).

[4] Russell, M. F., 'Automotive Diesel Engine Noise and its Control', SAE Paper 730243, *Society of Automotive Engineers Congress*, Detroit, Jan. (1973).

[5] Russell, M. F. and Cavanagh, E. J., 'Establishing Targets for Diesel Combustion Noise', SAE Paper 790271 in 80, *Society of Automotive Engineers Congress*, Detroit (1979).

[6] Russell, M. F., 'Recent CAV Research in Noise, Emissions and Fuel Economy of Diesel Engines', SAE Paper 770257, *Society of Automotive Engineers Congress*, Detroit, Feb. (1977) (Also published in SAE publication PT17 'The Measurement and Control of Particulate Emissions', pp. 161—176.)

[7] Meurer, J. S., 'The M-Combustion System of MAN — Evaluation of Reaction Kinetics Eliminates Knock', *SAE Trans.*, **64** (1956), pp. 250—257.

[8] Alcock, J. F. and Scott, W. M., 'Some More Light on Diesel Combustion', *Proc. I. Mech. E.*, (1962—3), No. 5 pp. 179—200.

[9] Khan, I., Greeves, G., Russell, M. F. and Warner, P. D., 'Prediction of Diesel Engine Noise', *I. Mech. E. Conference Land Transport Engines, Economics versus Environment*, Jan (1977), paper C15/77, pp. 139—154.

[10] Andree, A. and Pachernegg, S. J., 'Ignition Conditions in Diesel Engines', SAE Paper 690253, *Society of Automotive Engineers Congress*, Detroit Jan. (1969).

[11] Lyn, W. T. and Valdmanis, E., 'The Effects of Physical Factors on Ignition Delay', *Proc. I. Mech. E.*, **181** (1966—7), Part 2A.

[12] Austen, A. E. W. and Priede, T., 'Noise of Automotive Diesel Engines: Its Causes and Reduction', SAE Paper 1000A, *Society of Automotive Engineers International Automotive Engineers Congress,* Detroit, Jan. (1965).

[13] Priede, T., 'Relation between Form of Cylinder Pressure Diagram and Noise in Diesel Engines', Four Papers on Diesel Engine Fuel Injection, Combustion and Noise, *Proc. I. Mech. E.* (Auto Div.) (1960–61), No. 1, 63.

[14] Priede, T., Austen, A. E. W. and Grover, E. C., 'Effect of Engine Structure on Noise of Diesel Engines', *Proc. I. Mech. E.,* **179** (1964–5), Part 2A, No. 4, pp. 113–143.

[15] Anderton, D., 'Relation between Combustion System and Engine Noise', SAE Paper 790279, in P80 *Society of Automotive Engineers Congress,* Detroit (1979).

[16] Atkins, K. A. and Challen, B. J., 'A Practical Approach to Truck Noise Reduction', Appendix 1 of *I. Mech. E. Paper 131/79.*

[17] Head, H. E. and Wake, J. D., 'Noise of Diesel Engines under Transient Conditions', SAE Paper 800404, *Society of Automotive Engineers Congress,* Detroit, Feb. (1980).

[18] Hayes, P. A., Seybert, A. F. and Hamilton, J. F., 'A Coherent Model for Piston Impact Generated Noise', SAE Paper 790274, in P80 *Society of Automotive Engineers Congress,* Detroit (1979).

[19] Oetting, H. and Papez, S., 'Reducing Diesel Knock by Exhaust Gas Recirculation', SAE Paper 790268, in P80 *Society of Automotive Engineera Congress,* Detroit (1979).

[20] Rohrle, M. D., 'Affecting Diesel Engine Noise by the Piston', SAE Paper 750799, in SP397 *Society of Automotive Diesel Engine Noise Conference,* Milwaukee, Sept. (1975).

[21] Fawcett, J. N. and Burdess, J. S., 'Controlling Piston Slap', *Eng. Materials and Design,* March (1973), pp. 17–19.

[22] Haddad, S. D., 'Some Methods of Controlling Impact Noise in Engines', *Internoise 77,* Zurich, March (1977).

[23] Munro, R. and Parker, A., 'Transverse Movement Analysis and its Influence on Diesel Piston Design', SAE Paper 750800, in SP397, *Society of Automotive Engineers Diesel Engine Noise Conference,* Milwaukee, Sept. (1975).

[24] Ungar, E. E. and Ross, D., 'Piston Slap as a Source of Engine Noise', *ASME Paper 65-OGP-10* (1965).

[25] Zinchenko, V. I., *Noise of Marine Diesel Engines,* Sudpromigiz, 1957.

[26] Priede, T., Grover, E. C. and Lalor, N., 'Relation between Noise and Basic Structural Vibration of Diesel Engines', SAE Paper 690450, *Society of Automotive Engineers Congress,* Chicago, May (1969).

[27] Lalor, N., 'Computer Optimised Design of Engine Structures for Low Noise', SAE Paper 790364, *Society of Automotive Engineers Congress Detroit,* Jan. (1973).

[28] Russell, M. F., 'Diesel Engine Noise: Control at Source', SAE Paper 820238, *Society of Automotive Engineers Congress,* Detroit, March (1982).

[29] Thien, G. E., 'Use of Specially Designed Covers and Shields to Reduce Diesel Engine Noise', SAE Paper 730244, *Society of Automotive Engineers Congress,* Detroit, Jan. (1973).

8

Mechanically-induced noise and vibration in diesel engines

S. D. Haddad, Department of Engineering Technology, Western Michigan University

1. INTRODUCTION

Since the onset of the energy crisis, there has been an upward trend to use the diesel engine for light-duty vehicles and passenger cars in addition to its ever-increasing use in commercial automotive vehicles. This is mainly because the diesel engine is more efficient than its gasoline counterpart. However, the diesel engine operates at higher peak pressures, higher rates of pressure rise, and greater piston sideways forces, resulting in noisier operation, hence, the urgency to make it quieter or more socially acceptable.

In normally aspirated engines, the characteristic combustion explosion is usually the predominant source of excitation. Extensive investigation during the last decade has clarified the methods to control this combustion source, such as combustion chamber design, injection system layout, fumigation, pilot injection, and turbocharging.

Turbocharging has become fairly well established in the commercial vehicle automotive market mainly because of increased output, and this trend is expected to continue. Therefore, in the modern diesel engine, excitation from combustion is of secondary importance until such time as other sources have been substantially reduced. In fact, even for some normally aspirated engines, the mechanically-induced noise tends to predominate, especially at lower loads. This chapter basically summarizes results of research and development conducted by the author and his associates. The list of references (1 to 20) is provided for further information and guidance.

2. MECHANICALLY-INDUCED EXCITATION IN THE DIESEL ENGINE

In any reciprocating engine, certain clearances must be maintained between the running parts, as in the case of the piston in the bore, journal in the bearings, mating gears and fuel injection systems. In all these moving parts, the gas forces which act on the piston are modified by alternating forces generated in the crank mechanism. The result is that the force between the two moving parts, or between the moving and stationary parts, changes its direction or sign. During the operation of the engine, the moving parts are accelerated across the clearances, thus causing either mechanical impacts or impulsive hydraulic pressures in the lubricating oil film, causing subsequent structural deformation of the engine components.

The following are major sources of mechanically-induced noise and vibration in diesel engines:

- Piston slap — either mechanical impacts or impulsive oil film forces
- Timing gear impacts. — mechanical impact forces
- Bearing impacts — impulsive oil film forces
- Fuel injection system — hydraulic and mechanical forces
- Valve system and accessories — lubricating oil pump, compressors when fitted, the turbocharger unit itself, etc.

3. PISTON SLAP AS A MAJOR SOURCE OF DIESEL ENGINE NOISE AND VIBRATION

Since 1968 (at VÚNM, ISVR, Lucas CAV and LUT)*, extensive investigations have been pursued to provide a better understanding of the mechanism of mechanical noise generation and to recommend appropriate methods of noise control.

In the course of these investigations it has been shown that piston slap excitation is the most important mechanical source of noise and vibration in the diesel engine. The following methods were used to quantify and provide an in-depth study of piston slap:

1. Theoretical treatment.
2. Experimental simulation.
3. Analogue simulation.
4. Studies on running engines.

3.1 Theoretical treatment of piston behaviour in a diesel engine
Piston slap is initiated whenever the piston side thrust force changes direction. This takes place under two conditions: (a) when the force in the connecting rod changes from tension to compression and vice versa and (b) when the component of the connecting rod forces normal to the cylinder axis changes

* VÚNM: Oil Engine Research Institute — Prague.
  ISVR: Institute of Sound and Vibration Research — Southampton University.
  Lucas CAV Ltd., Acton, London.
  LUT: Loughborough University of Technology.

direction as a result of changes in crank angle. The latter condition always occurs at top and bottom dead centre whereas the former condition is realized when the total inertia force contribution to side thrust just balances the resulting gas force. In conventional diesel (or petrol) engines, it is always found that piston slap occurs at or near the top and bottom dead centres while mid-stroke slaps may be suppressed at low speed operating condition when the aforementioned force balance cannot be obtained. Every one of these impacts will generate a certain amount of kinetic energy which is imparted to the liner. This energy is transmitted through the engine structure and modified owing to elasticity, damping, etc., to be radiated in the form of surface vibration and engine noise. The kinetic energy may be calculated from a basic formulation of the piston connecting rod dynamics, for which a computer program was written and developed at Wellworthy Ltd., in liaison with the author from 1972–75 and has since been appreciably updated and used to predict the optimum piston design features to minimize noise from piston slap in a numb er of diesel engines.

3.1.1 Description of the Basic Program
Ther program is written in FORTRAN IV and uses certain simplifying assumptions which are listed below:

(1) The piston contacts the cylinder bore only at the top and/or bottom of the skirt.
(2) The piston crown does not touch the bore at any stage.
(3) All impact forces are instantaneous.
(4) Skirt and liner elasticity are neglected and so no account is taken of piston rebound after impact.
(5) The piston skirt/bore oil film has little effect on the transverse piston motion.
(6) The force due to the gas pressure acts through the centreline of the piston, rather than the centreline of the bore, so that the equations of motion apply no matter where the piston is in the bore.

At each increment of the crank angle it calculates the piston-to-bore side thrust at the top and bottom of the skirt, and monitors the direction of these forces. If one (or both) changes direction, then the basis of the calculation is altered and the mode of motion changes. The four basic modes are shown in Fig. 1.

In modes 2, 3 and 4, the results are usually printed out every 1° of crank angle for four strokes and every ½° for two strokes, but these increments may be changed to suit any application. In mode 1, the results are printed out every 10° crank angle. Position indicators for the top and bo ttom of the skirt show with which side of the bore the piston is in contact, or was last in contact. When the calculation of lateral piston position indicates that the clearance has been taken up at the top and/or bottom of the skirt, then the program changes mode and calculates the kinetic energy released at impact.

MODE 1 Both top and bottom of skirt are in contact
with the bore, either vertically or diagonally

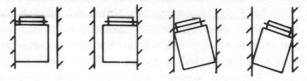

MODE 2 Top of skirt in contact with one side of bore

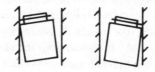

MODE 3 Bottom of skirt in contact with one side of bore

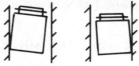

MODE 4 Piston free in bore

Fig. 1 – Basic modes of motion of the piston in the bore.

The instantaneously released kinetic energy is calculated using the theory for conservation of moment of momentum, which states that the total moment of momentum, before impact, is equal to the total moment of momentum after impact, bearing in mind that both translational and rotational piston impact velocities are considered. The program then adds the instantaneous kinetic energies to produce the total kinetic energy released in one complete engine cycle (called the Cycle KE).

3.1.2 Kinetic Energy–Noise Relationship
A number of researchers have shown that the engine block is basically a linear structure and that a near-linear relationship exists between engine noise and engine block response. Therefore, it can be deduced that a near-linear relation-

ship should be obtained between the total kinetic energy due to piston impact excitation and the resultant engine noise (remembering that piston slap is assumed predominant here). Experimental work by the author has shown this relationship to hold for a number of diesel engines and this has established confidence in the predictive ability of the program (in spite of the assumptions) and also in the logic of the kinetic energy versus noise concept. The program has since been used to predict the effect of piston design on noise level.

3.1.3 Low Noise versus Mechanical Efficiency

When looking at ways of reducing the mechanical noise of an engine it is important to consider the effect on mechanical efficiency, which is related to frictional losses in the engine. For example, by using smaller piston/bore clearances the severity of piston impacts is reduced, but this is at the expense of increased friction and oil drag and therefore lost efficiency. When supplied with the appropriate friction coefficients, the piston motion program computes, at intervals of $1°$ crank angle over a complete cycle, the frictional forces acting between the piston and the bore, the piston boss and the gudgeon pin and between the top ring and the groove. Now the total friction mean effective pressure (f.m.p.) of the engine includes not only these three components but also the friction due to other moving parts in the engine, the pumping work and also the power required to drive auxiliary equipment.

From a survey of relevant data on this subject it was possible to derive a speed-dependent factor which accounts for the other components to give an estimate of the total f.m.e.p., which leads to an estimate of the mechanical efficiency η_m (η_m = i.m.e.p. − f.m.e.p./i.m.e.p.). Therefore, from an input consisting of information about the piston/connecting rod system, engine speed, measured cylinder pressure diagrams and values of the frictional coefficients, the present piston motion program produces the following:

(1) the bore clearances at the top and bottom of the skirt as the piston moves across the bore;
(2) the translational and impact velocities at the top and bottom of the skirt;
(3) the attitude of the piston to the bore;
(4) the piston-to-bore side thrust at the top and bottom of the skirt and the gudgeon pin side thrust;
(5) the points in the engine cycle at which lateral piston movement and piston slap impact occur;
(6) the instantaneous kinetic energy imparted to the engine liner at each piston/bore impact;
(7) the total kinetic energy during one cycle of engine operation;
(8) the estimated mechanical efficiency of the engine.

Figure 2 shows a specimen output, all of which can be produced in the form of plots against crank angle for ease of evaluation.

```
1645-12E3B-G-M-ENGINE
 FULL LOAD
 PISTON DIAMETER                         =   2.2987E-01 M
 WEIGHT OF PISTON AND RINGS              =   2.7579E 02 N
 DIST TOP OF SKIRT TO G.P. CENTRE        =   1.0348E-01 M
 DIST BOT OF SKIRT TO G.P. CENTRE        =   1.3945E-01 M
 PISTON/BORE CLEARANCE TOP OF SKIRT=         6.3500E-04 M
 PISTON/BORE CLEARANCE BOT OF SKIRT=         2.2860E-04 M
 CRANK OFFSET FROM BORE CENTRE LINE      =   0.0000E-01 M
 G.P.OFFSET FROM BORE CENTRE LINE        =   0.0000E-01 M
 PISTON C.G OFFSET FROM BORE C/L         =   0.0000E-01 M
 TIME FOR PISTON SWINGS                  =   57.1 SECS
 NUMBER OF PISTON SWINGS                 =   50.0
 DIST FROM PISTON C.G TO G.P C/L         =   4.3942E-02 M
 DISTANCE G.P. C/L TO TOP RING           =   1.4303E-01 M
 ANGLE OF TOP RING                       =   0.0000E-01 DEGS
 BORE DIAMETER                           =   2.3011E-01 M
 TOP RING RADIAL WIDTH                   =   7.6200E-03 M

 WEIGHT OF CON ROD AND GUDGEON PIN=         3.7089E 02 N
 CON ROD LENGTH CENTRES                  =   5.8420E-01 M
 DIST CON ROD C.G TO SMALL END CENTRE=       4.0558E-01 M
 CRANK RADIUS                            =   1.2700E-01 M
 BIG END DIAMETER                        =   1.6510E-01 M
 TIME FOR CON ROD SWINGS                 =   53.2 SECS
 NUMBER OF CON ROD SWINGS                =   50.0
 GUDGEON PIN DIAMETER                    =   9.3574E-02 M
 DIST BOTTOM OF LINER TO CRANK C/L       =   3.2225E-01 M

 COEFF OF FRICT PISTON TO G.F.=    0.010
 COEFF OF FRICT PISTON TO LINER =  0.010
 COEFF OF FRICT TOP RING TO PISTON =      2.0000E-01
 ENGINE SPEED                            =   900.0 R.P.M.

 GAS PRESSURES N/SQ M  0-385 DEGS  5.0 DEG INTERVALS

 MODE1=NO LATERAL PISTON MOVEMENT
 MODE2=PISTON ROTATION ABOUT TOP OF SKIRT
 MODE3=PISTON ROTATION ABOUT BOTTOM OF SKIRT
 MODE4=PISTON FREE IN BORE
 SKIRT CLEARANCE MEASURED ON NON-THRUST SIDE
```

ENGINE SPEED TO BE VARIED FROM 450.0 TO 900.0 IN INCREMENTS OF 50.0

1645-12E3B-G.M-ENGINE

		SKIRT CLEARANCE		SKIRT VELOCITY		PISTON TO BORE ATTITUDE	PISTON SIDE THRUST			KINETIC ENERGY LOSS AT IMPACT
MODE	THETA (DEGREES)	TOP X1 (N)	BOTTOM X2 (N)	TOP U1 (M/SEC)	BOTTOM U2 (M/SEC)	PHI D (DEGREES)	TOP F1 (N)	BOTTOM F2 (N)	TOTAL A2 (N)	KEL (N-M)
1	20.0	(0.000E-01)	0.000E-01	0.000E-01	0.000E-01	-0.048	742.1	561.5	-1303.6	0.00
1	30.0						1000.3	756.8	-1757.0	0.00
1	40.0						1142.3	864.2	-2006.6	
1	50.0						1158.7	876.6	-2035.4	
1	60.0						1058.9	801.1	-1860.0	
1	70.0						869.2	657.6	-1526.9	
1	80.0						627.2	474.5	-1101.7	
1	90.0						373.4	282.5	-656.0	
1	100.0						143.5	108.6	-252.1	
1	108.0	2.087E-08	1.513E-08	1.127E-04	8.172E-05	-0.048	0.0	0.0	8.0	0.00
4	109.0	1.253E-07	1.268E-07	4.515E-04	5.211E-04	-0.048	0.0	0.0	27.6	0.00
4	363.0	6.335E-04	2.263E-04	-6.849E-03	-1.048E-02	0.048	0.0	0.0	-586.5	0.00
4	364.0	6.318E-04	2.238E-04	-1.138E-02	-1.751E-02	0.048	0.0	0.0	-780.6	0.00
4	365.0	6.292E-04	2.197E-04	-1.703E-02	-2.631E-02	0.049	0.0	0.0	-977.8	0.00
4	366.0	6.254E-04	2.139E-04	-2.378E-02	-3.685E-02	0.049	0.0	0.0	-1165.8	0.00
4	367.0	6.203E-04	2.059E-04	-3.164E-02	-4.912E-02	0.050	0.0	0.0	-1356.3	0.00
4	368.0	6.136E-04	1.955E-04	-4.058E-02	-6.311E-02	0.051	0.0	0.0	-1545.2	0.00
4	369.0	6.052E-04	1.824E-04	-5.061E-02	-7.881E-02	0.052	0.0	0.0	-1732.2	0.00
4	370.0	5.948E-04	1.662E-04	-6.170E-02	-9.619E-02	0.053	0.0	0.0	-1917.1	0.00
4	371.0	5.822E-04	1.466E-04	-7.385E-02	-1.152E-01	0.055	0.0	0.0	-2099.5	0.00
4	372.0	5.673E-04	1.233E-04	-8.703E-02	-1.359E-01	0.057	0.0	0.0	-2279.5	0.00
4	373.0	5.499E-04	9.609E-05	-1.012E-01	-1.582E-01	0.059	0.0	0.0	-2456.7	0.00
4	374.0	5.297E-04	6.458E-05	-1.165E-01	-1.821E-01	0.062	0.0	0.0	-2630.8	0.00
4	375.0	5.067E-04	2.851E-05	-1.327E-01	-2.075E-01	0.065	0.0	0.0	-2801.7	0.00
3	376.0	4.775E-04	0.000E-01	-1.066E-01	-2.263E-01	0.065	0.0	2016.7	-3757.3	0.43
3	377.0	4.563E-04	0.000E-01	-1.227E-01	-2.263E-01	0.060	0.0	2126.5	-3964.4	0.00
3	378.0	4.320E-04	0.000E-01	-1.395E-01	0.000E-01	0.054	0.0	2233.6	-4166.6	0.00
3	379.0	4.045E-04	0.000E-01	-1.572E-01	0.000E-01	0.047	0.0	2338.0	-4363.8	0.00
3	380.0	3.737E-04	0.000E-01	-1.757E-01	0.000E-01	0.040	0.0	2439.6	-4555.7	0.00

TOTAL KINETIC ENERGY LOSS AT IMPACT IN CYCLE = 5.672 J

PISTON FMEP = .16206E 03 N/SQ. M.

MECHANICAL EFFICIENCY OVER CYCLE = 99.3 PER CENT

Fig. 2 – Part of output from the piston motion program.

The foregoing analytical treatment embodied in the piston motion computer program has been used to predict the optimum piston design features for low piston-slap-induced noise and acceptable mechanical efficiency. Figure 3 shows an example of this study which was used to predict the optimum gudgeon pin offset piston for low noise whereas Fig. 4 presents two examples of the trade-off concept for gudgeon pin (GP) and centre of gravity (CG) offset pistons.

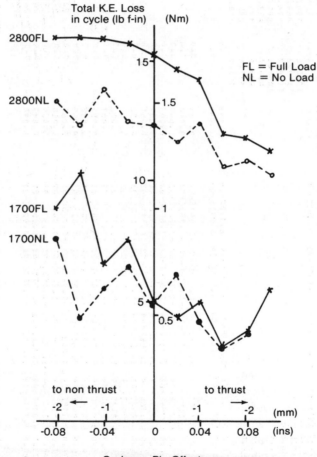

Fig. 3 – Effect of gudgeon pin offset on total kinetic energy loss of piston at impact.

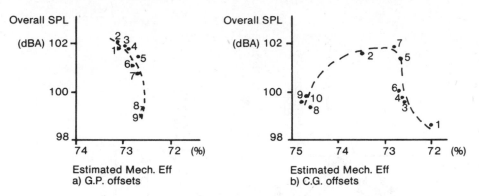

Estimated Mech. Eff
a) G.P. offsets

Estimated Mech. Eff
b) C.G. offsets

NB. Numbers 1-10 represent offsets from -2.032mm (non thrust side)
to 2.54mm (thrust side) in steps of 0.508mm
Zero offset is therefore number 5.

Fig. 4 – Effect of gudgeon pin offset and centre of gravity offset on overall SPL
and estimated mechanical efficiency (at 2800 rev/min full load).

3.2 Experimental Simulation of Piston Slap

Following the need to clarify the contribution of piston slap excitation to overall engine noise, the author developed a piston slap simulation technique for a V8 diesel engine as shown in Fig. 5. In this technique, a hydraulic force generator applies the force through a monitoring force transducer to the piston via twin back-to-back connecting rods using the crankshaft as a pivot. The crankshaft bearings are packed with thick grease or a thin rubber sheet to prevent distortion of the applied force due to possible impacts which may occur in the plain bearings. A special oscillator is used to reproduce the calculated sideways force of the piston. The piston can then be excited at frequencies corresponding to any engine speed and the response measured to evaluate piston slap excitation. The results of this study have shown that:

(1) the rate of rise of piston sideways force (K) is one of the most important parameters controlling piston slap-induced engine noise and vibration;

(2) the rig can be relied on as a tool for designers to study engine vibration and noise response at an early stage, even before assembling and running the engine in its final form;

(3) this technique can be adapted to simulate bearing impacts and gear rattle to estimate their individual contributions;

(4) the effect of a number of piston design features tested in the rig correlated well with the corresponding effect in the running engine, after allowing for certain corrections. These design features included the effect of piston—liner clearance, oil film thickness and viscosity, change of piston mass and change of slap incidence.

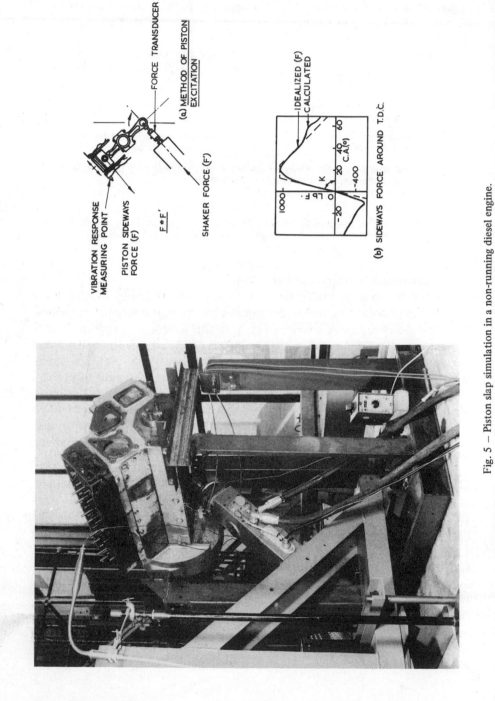

Fig. 5 — Piston slap simulation in a non-running diesel engine.

The next stage was to develop the present system for an in-line engine for similar studies of piston slap excitation. In the process, certain improvements were introduced to make the simulation even more realistic than in the case of the vee-type rig.

Figure 6 shows the general layout of the in-line piston slap simulation rig. In this case, because of engine size and characteristic engine structure frequencies

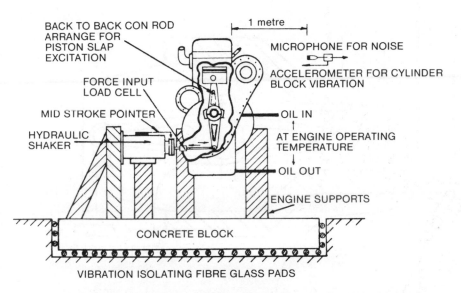

Fig. 6 — Layout of simulation rig for the study of piston slap excitation in an in-line, four-cylinder diesel engine.

it was found difficult to design the engine support frame to avoid interference with any exciting frequencies. Therefore a hydraulically and pneumatically operated system shown in Fig. 7 was subsequently developed and used successfully for parametric study of piston slap-induced noise and vibration in in-line diesel engines.

3.3 Analogue Simulation of Piston Slap

For this purpose a simplified mathematical model of the piston—liner assembly was formulated as shown in Fig. 8. Impact is modelled as between two elastic bodies with coefficient of restitution e.

An analogue EAI 580 computer was used for the simulation of this model. This had logic control of analogue computer modes and various facilities which enabled the effect of various changes in the system to be studied. The results from this study, in spite of the simplified model, produced useful trend information in good agreement with the previous techniques.

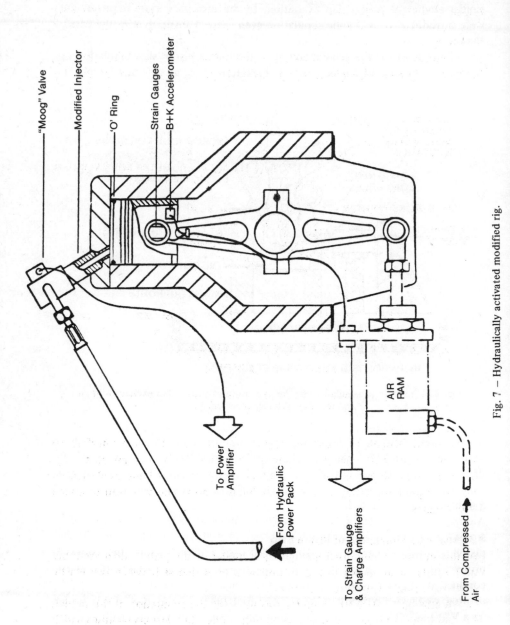

"Moog" Valve

Modified Injector

'O' Ring

Strain Gauges

B+K Accelerometer

To Power Amplifier

From Hydraulic Power Pack

AIR RAM

To Strain Gauge & Charge Amplifiers

From Compressed Air

Fig. 7 – Hydraulically activated modified rig.

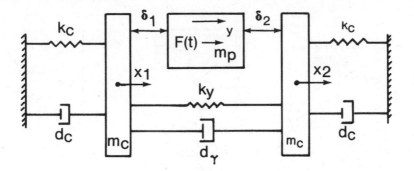

Fig. 8 – Simplified model for piston–liner movement near TDC.

From these studies it was found that, in addition to other parameters, the piston impact velocity (and hence the KE imparted to the liner) around top dead centre (TDC) is directly related to engine block vibration and the resultant engine noise. In its simplified form the impact velocity (V_i)

$$V_i \propto \left(\frac{18K\delta^2}{m_p}\right)^{\frac{1}{3}}$$

and the kinetic energy in the piston at impact is thus (KE_i)

$$KE_i \propto 3.43(m_p K^2 \; \delta^4)^{\frac{1}{3}}$$

where
$\quad K$ = rate of change of piston sideways force, dF/dt, at TDC
$\quad 2\,\delta$ = the liner–piston nominal clearance
$\quad m_p$ = piston mass

That is, for a certain piston–liner design, the rate of rise of piston sideways force (K) is the controlling parameter.

3.4 Studies on Running Engines
Alongside the analytical treatment and the simulation techniques of piston slap, studies were conducted on running engines to evaluate the effect of changing piston design features on overall engine noise. Figure 9 shows the effect of progressive change in piston-to-bore clearance on overall engine noise when conducted on a multi-cylinder, in-line diesel engine. Figure 10 shows the effect of progressive change in GP offset pistons on overall engine noise when conducted on a V8 diesel engine.

In these tests it is important to ensure that the combustion characteristics have not changed in order to be able to arrive at the pure effect of piston slap. This is not always easy to do.

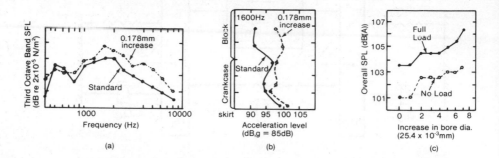

(a)   (b)   (c)

Fig. 9 – Noise and vibration characteristics with changing piston-to-bore clearance in a multi-cylinder, in-line diesel engine. Engine B: six cylinders. (a) Noise spectra showing effect of increased bore diameter; 2800 rev/min, full load; (b) Vibration pattern down exhaust side of engine; 2800 rev/min, full load; (c) Effect of increased bore diameter on overall noise; 2800 rev/min, SPL = sound pressure level at 1 m from engine level with exhaust manifolds. 'Standard' piston-to-bore clearance = 0.254 mm; '0.178 mm increase' piston-to-bore clearance = 0.432 mm.

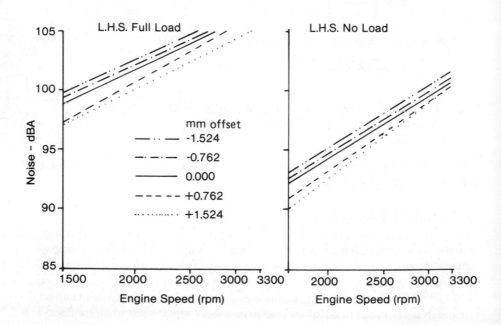

Fig. 10 – Effect of GP offset pistons on overall engine noise at 1 m from thrust side in a V8 automotive diesel engine.

To identify the contribution of piston slap to overall noise in a running engine, an oscillographic technique is used. This technique enables the separation of individual events in the engine cycle and is outlined in Fig. 11. For this, three signals are needed:

1. Degree marks to locate the sequence of firing cylinders; for this a degree marker disc is fitted to the front of the crankshaft to indicate $10°$ intervals and the cylinder TDC position. A special magnetic pick-up is used to feed the degree marker signal into one channel of an oscilloscope.
2. Pressure diagram (Fig. 11(a)) to locate the reference firing cylinder; for this a quartz transducer is installed in the cylinder. This is positioned such that the diaphragm is flush with the surface of the combustion chamber; the signal is fed through a charge amplifier into a second channel of the oscilloscope.
3. Either noise or vibration (Fig. 11(c)) of the engine; noise and vibration are measured using Bruel and Kjaer microphones, accelerometers and associated

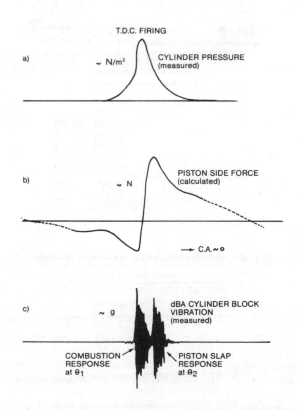

Fig. 11 – The Oscillographic Method.

equipment. Before display onto a third channel of the oscilloscope, either signal is fed through a spectrometer set on the dB(A) weighting scale. The dB(A) scale is found most representative for this comparative study.

In addition to the above signals, the resultant piston side force is calculated (Fig. 11(b)).

As shown in Fig. 11(c), two response signals may be distinguished: the first coincides with the start of combustion at crank angle θ_1 before top dead centre (BTDC) and the second occurs at θ_2 after top dead centre (ATDC) around the

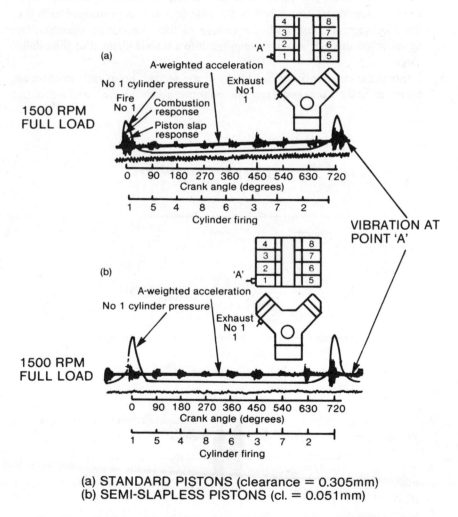

(a) STANDARD PISTONS (clearance = 0.305mm)
(b) SEMI-SLAPLESS PISTONS (cl. = 0.051mm)

Fig. 12 — Oscillographic identification of piston slap-induced engine structure vibration.

point where piston slap is expected to occur. In some engines and at certain conditions, these two major response pulses may overlap owing to close incidence. There are two methods to facilitate a separation:

(i) Use a drum camera to expand the time base. This can also be accomplished by photographic enlargement of single-shot polaroid pictures.
(ii) Tape record the signals and use a data analysis facility to separate the two contributions.

Figure 12 shows an example of using the oscillographic technique to evaluate the effect of piston-to-liner clearance on the resultant engine vibration response.

Finally, radio telemetry, slip rings or a linkage can be equally useful to measure actual piston behaviour in a running diesel engine and therefore be able to quantify piston slap-induced noise and vibration. This line of research has shown the importance of the thin lubricating oil film pressures entrapped between the piston and the liner, especially due to the major piston slap around TDC, which can deform the cylinder liner and subsequently the engine block. The resultant response was shown to be an important exciting force controlling engine block vibration and noise in the frequency range 1000 Hz upwards.

4. METHODS OF REDUCING PISTON SLAP-INDUCED NOISE AND VIBRATION IN DIESEL ENGINES

The above studies have clarified the mechanism of piston slap noise generation leading to some of the following methods to control piston slap noise and vibration:

1. Offsetting the gudgeon pin of the piston *normally* by around 0.060 in. to the thrust side depending on engine size, combustion pressures, timing and piston CG (in rare cases offsetting to non-thrust side).
2. Reducing the piston–liner clearance; in this case $20 \log_{10} C_1/C_2$ must be greater than 4 dB for measurable noise reductions to be obtained (C_1, C_2 being the upper and lower clearances, respectively). This method can be applied by using expansion-restricted (strut) pistons or special piston skirt surface treatment using PTFE oil control rings.
3. Using articulated pistons (cross-head); the top half, which carries the combustion load, consists of the piston crown with two trunnions, each with holes to accommodate the wrist pin.
4. Modifying the combustion characteristics, that is the injection timing and inlet spray. This is a more complicated process which usually requires close study of the engine under investigation.
5. Lengthening the piston skirt (with an extremely short skirt and a large amount of play, the inclination of the piston in the cylinder is considerable).

6. The piston should be of maximum diameter and of sufficiently rigid shape not to become deformed under the mechanical stresses occurring in service.

7. The liner should be rigid and stresses arising in assembly should be relieved to prevent very small deformations or 'ovality'.

8. There should be overall low clearances for the piston pin and bearings.

9. By suitable introduction of coolant, care should be taken to have the minimum possible differences in temperature between both the upper and lower ends of the cylinder. With uneven cooling, the cylinder expands funnel-shaped against the combustion chamber. The piston may then tend to stick in the lower half of the cylinder whereas in passing TDC it will have excessive play.

10. Introduction of oil cushions between the piston and the cylinder wall, e.g. as obtained by forcing oil through ports in the piston wall between two rings or using special two-scraper rings in the skirt of the piston. Also using resilient piston skirts.

11. Isolation of the cylinder liner from the engine block and possibly increasing the damping of the liner.

12. Using crankshafts offset from the cylinder centreline in the direction to delay piston slap at TDC.

13. The ratio of crank radius to connecting rod length should be kept small.

14. Using methods of holding pistons to one wall of the liner and rearranging pressure-feed lubrication, e.g. elliptical piston rings at an angle less than $90°$ to the piston axis, sprung shoe incorporated in skirt and development of hydrostatic pressure.

5. OTHER SOURCES OF MECHANICALLY-INDUCED NOISE AND VIBRATION IN DIESEL ENGINES AND THEIR METHODS OF CONTROL

An engine running without any gas loading on a piston will produce a certain level of noise owing to various sources of a mechanical nature. Any application of gas load on a piston therefore introduces another source which increases the resulting emitted noise.

The summation of these two noise sources depends on their relative levels, the relation being illustrated in Fig. 13.

Assuming the level of mechanical noise of the engine to be 100 dB, the effect of the noise produced by gas loads will be just noticeable (*increase of 0.4 dB*) when this noise is about *10 dB below the mechanical noise*. When the noise due to gas loads is equal to the mechanical noise, the total level of the noise is increased by 3 dB. A further increase of the noise due to gas loads increases the total noise until a condition is reached when noise is fully controlled by the gas loads.

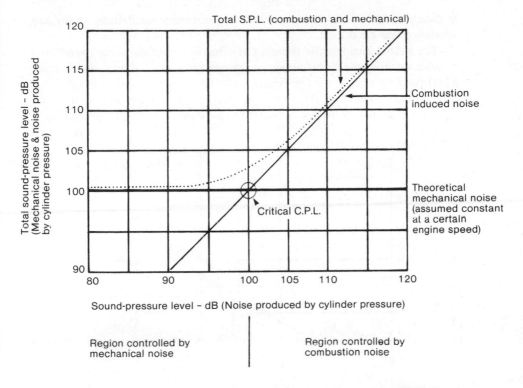

Fig. 13 — Summation of mechanically and combustion-induced noise at a certain engine speed.

The condition when noise due to combustion has attained the same level as the general mechanical noise is termed as the 'critical cylinder pressure level' of the engine.

Figure 13 can be applied for the addition of any two important noise sources. If combustion-induced noise is reduced then piston slap-induced noise would become predominant. Subsequently, if piston slap-induced noise is optimized then usually timing gear noise would predominate in diesel engines. The same argument applies to bearing impact noise, fuel injection equipment noise and other mechanically-induced noise sources in the engine.

6. TIMING GEAR NOISE AND ITS CONTROL

Vibrations and noise in gear arrangements are caused by various fluctuating forces present in the system. A constant force will not produce noise. Backlash in mating gears tends to accentuate the resulting mechanical impact forces between contacting teeth. The impacts between the gear teeth can be produced by changes in tangential force reinforced by torsional oscillations by bending vibration of the crankshaft.

The various engine noise sources can often be identified by the overall level of noise versus speed measurement, as shown for a four-cylinder diesel engine (Fig. 14).

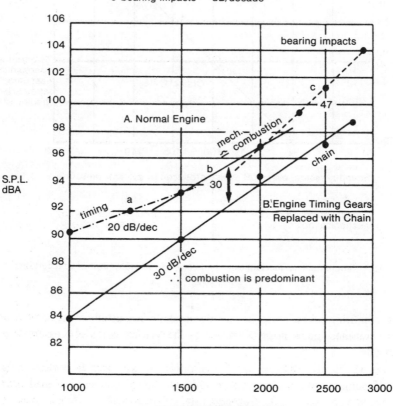

a timing gears.... due to meshing 20 dB/decade
b combustion control 30 dB/decade
c bearing impacts   dB/decade

Fig. 14 – Effect of timing drive on noise–speed relationship for a four-cylinder diesel engine.

Three well-defined regions can be seen in the overall noise level plot for a normal engine (Plot A):

(a) from 1000 to 1500 rev/min the slope is about 20 dB per tenfold increase of speed;

(b) from 1500 to 2000 rev/min the slope is 30 dB per decade;

(c) from 2000 rev/min upwards, the slope is 47 dB per decade.

This indicates that sources of considerably different nature influence the noise in various speed ranges.

The 20 dB slope is typical for timing gears where the loads are applied suddenly by the meshing of the individual teeth. This suddenness is generally accentuated by backlash. At the higher speeds the tooth meshing frequency becomes sufficiently high to coincide with major natural frequencies of the engine structure and thus a rapid increase of noise due to dynamic magnification may result at speeds exceeding 2000 rev/min. Plot B shows the effect of replacement of the timing gears with a chain drive and that the conclusions derived from the simple dB(A) measurements are valid. The 30 dB/decade slope suggests that combustion noise is now predominant and the reduction of 3 dB(A) in the 'b' range of the curve suggests that here combustion and mechanical noise were of about equal level.

It has been shown that to produce quieter gear operation, the designer should aim at methods leading to smoother application of contact forces (source attenuation) as in the case of double helical gears, changing the tooth profiles and replacement with chain or reinforced toothed belts. In a conventional diesel engine the timing gear system is located at the front of the engine and is invariably responsible for the predominant frontal noise which is largely influenced by gear clearance. Experiments involving replacement of the gears with a chain drive (or locating the gears at the rear of the engine) generally show some 4 to 6 dB(A) reduction of noise to the front and up to 3 dB(A) to the sides. Typical results of this investigation are shown in Figs. 15, 16 and 17 for a four-cylinder, in-line diesel engine.

The timing gears were replaced with a continuous chain drive system as shown in Fig. 15. Isolating the front timing cover would have been insufficient since any impact noise originating in the gears would be transmitted to the main engine structure and radiated, to a greater or lesser degree, by the whole engine structure. The cylinder pressure and injection timing were checked to be identical, before and after conversion. The effect of the modified timing system on the overall engine noise is shown in Fig. 16. Significant overall noise reductions are obtained to the front of the engine and throughout the speed range. At 1000 rev/min, the overall level is reduced by 6 dB(A) at full load and 4 dB(A) at no load. At higher engine speeds, the full load levels are reduced by 1.5 dB(A) at 1500 rev/min to 5 dB(A) at 2800 rev/min, whereas the no-load noise levels were down by 3–4 dB(A) at all speeds. Smaller overall noise reductions were obtained to the side of the engine.

Figure 17 presents typical third-octave noise spectra to show the effect of the conversion from gears to chain at 1000 rev/min. It can be seen that the gear impacts contribute to the overall engine noise in the acoustically important frequency range of 800 Hz and upwards. Somewhat less overall noise reductions were obtained on this engine by replacing the standard timing gears with double helical gears of optimum helix angles of 25°. Isolating the timing cover at the

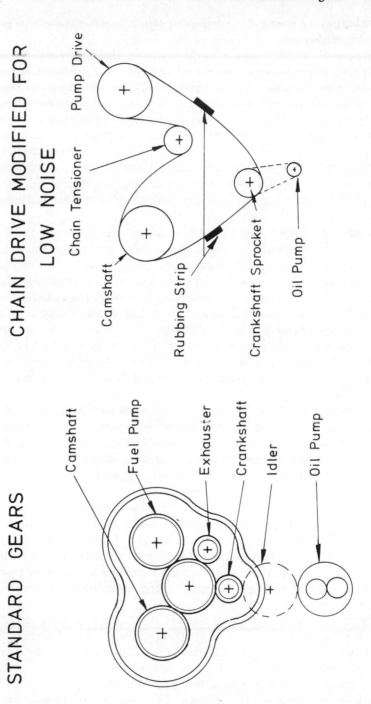

Fig. 15 — Timing gear and chain drive layout for diesel engine A.

mounting flange or treating it with suitable damping material (response attenuation) gives further reductions in engine noise.

    Reinforced toothed belts, if made to a high standard, can also usually
be used for automotive diesel engines and can give similar noise reduction to
that of the timing chain drives. On the other hand, the timing gear design can be
improved for quiet operation. A good summary of the way in which these gear
tooth design parameters — contact ratio, transmission error, load, speed, etc. —
affect noise is given by Opitz [17].

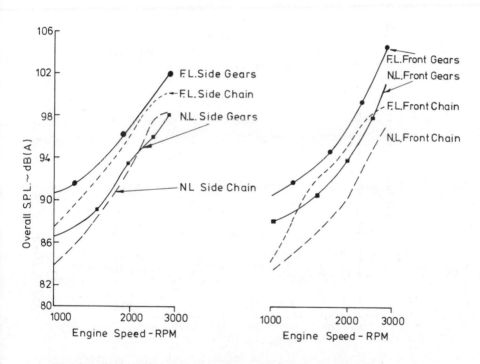

Fig. 16 — Effect on overall noise at 1 m due to replacing the timing gears with
chain drive on a four-cylinder, in-line diesel engine(A)

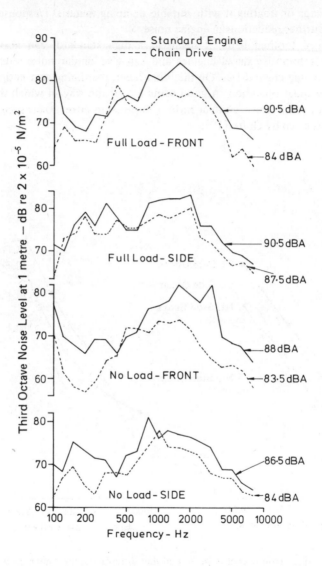

Fig. 17 – Typical spectra of noise reduction due to replacing the timing gears with chain drive at 1000 rev/min – engine A.

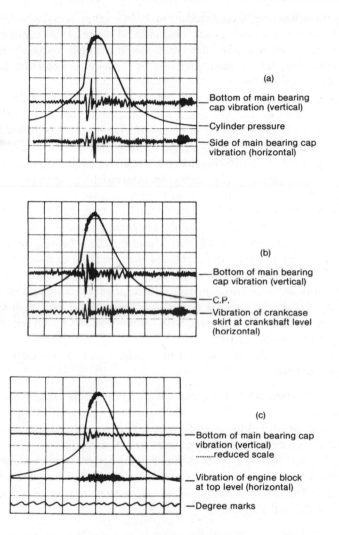

Fig. 18 — Oscillograms showing bearing impact vibration (dB(A) weighted) compared with block vibration corresponding to no. 6 cylinder centreline of six-cylinder, in-line diesel engine (C) at 1200 rev/min full load.

7. BEARING IMPACTS

Bearing noise has been less significant in diesel engines because of the predomin-ance of combustion and piston slap induced noise. In the gasoline engine, however, it has been shown that the predominant high frequency noise in the range 1000—3000 Hz is primarily induced by the impulsive pressure development in the lubricating oil film.

This same phenomenon is now being observed in some medium and small high-speed diesel engines. The operation of the crankshaft in the main bearings of an engine plays a vital role in determining structural excitation and is con-stituted by the form of the load applied to the engine structure via the oil film in the bearing clearance. Following this, it has been found that significant reductions in structurally radiated noise might be achieved in some types of engine by modifying the form and operation of the bearings in order to achieve smoother oil film pressure development. Around TDC, a large component of the com-bustion force is transmitted through the connecting rod to the bearings, initiating a sharp rise in the bearing oil film pressure called 'bearing impacts'. These pulses excite the journals, the crankcase skirt and the engine block to certain levels attenuated or amplified depending on the characteristics of the trans-mission paths and structure impedance, subsequently producing engine vibration and noise. Figure 18 shows a typical oscillographic investigation to illustrate bearing impact response compared with combustion and piston slap responses at three positions on the engine.

Some of the following methods have been found to give some reduction in bearing impact noise:

1. Specially developed bearing load computer programs to optimize the bearing forces.
2. In some small, lightly loaded engines it is possible to use special rubber-backed bearings or to introduce some sort of damping.
3. Appropriate form and force closure bearing designs to prevent impacts.
4. Reduction in the working clearances; but the use of extremely small clear-ances affects the reliability of the engine (seizure may occur) and reduces the mechanical efficiency.

8. NOISE OF FUEL INJECTION SYSTEMS

The main sources of noise in the fuel injection systems are:

1. The fuel pump generating hydraulic forces of near-sawtooth-like nature with typical response at around 20 dB/decade.
2. Injectors generating needle-seating impacts where the force is provided via a spring. In this case the impulses are independent of the number of injections per unit time and the resultant noise remains constant with engine speed.

3. Fuel cavitation; if severe, this can become the predominant source of noise as well as inducing harmful side-effects with regard to the efficient running of the fuel system.

Fuel injection systems vary in type, design and their contribution to the overall engine noise. Generally, on present generation diesel engines, they pose little problem but might require attention in future low noise units.

The results of an investigation are described in brief showing the contribution to overall engine noise of one type of fuel injection system installed on a V8 engine. The fuel injection equipment was motored independently on the engine by means of an electric motor. The engine crankshaft was replaced with one without a driver gear. The air compressor was removed and the fuel pump directly connected through the camshaft gear and motored by means of a double belt drive from the electric motor, as shown in Fig. 19. The tappets were held out of contact with the cams, and the fuel was collected through the exhaust manifolds. An external oil pump was used to circulate the lubricating oil.

With the fuel injection equipment motored, noise was measured at 1 m from the engine sides, top and front for direct comparison with that of the fully-running engine. Varying load conditions were obtained by adjusting the fuel rack. Figure 20 shows typical comparisons of third-octave noise spectra for this type of motored fuel injection system with engine speed at full load conditions and only for right-hand side and top of engine. Noise spectra of the fully running engine are also shown. It can be seen that this fuel injection system has much lower noise levels than that of the complete engine. The peaks in the noise spectra of the motored fuel system around 1000–1250 Hz correspond to the fuel pump gear tooth meshing frequencies, suggesting that a change in tooth meshing should reduce these peaks. Detailed third-octave vibration analysis was also conducted together with an exercise of covering the various parts with 1 in. fibreglass and lead, confirming that the fuel pump is the main noise source of this fuel injection system.

The following are some of the methods used to control fuel injection system noise:

(a) The fuel pump should be treated in the same way as the engine structure, that is a stiffer structure with a certain amount of damping. Also, newly developed rotary fuel injection pumps are inherently quiet and should give large reductions in overall noise levels of the total fuel injection system, and subsequently in the engine noise.

(b) Attention to camshaft design to ensure smoother operation.

(c) Use of injector needles of damped construction and of spring-loaded parts, in general.

(d) Minimization of the effect of impacts by reducing impact forces or adjusting impact incidence.

External Lubricating oil system

Fig. 20 – FIE noise contribution on diesel engine B.

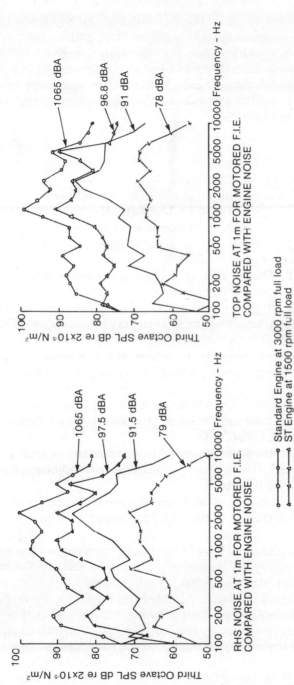

Fig. 19 – Fuel injection equipment motored (engine B).

Most of the noise control methods discussed above can be adapted to reduce the noise of valve mechanisms, accessories and other engine noise source that becomes either equal to or higher than any other noise component. If the engine designer is permitted to introduce most or all of the above effective techniques to optimize combustion-induced and mechanically-induced noise sources, it should be possible to produce low noise diesel engines without having to use any palliative measures or enclosures.

REFERENCES

[1] Hempel, W., 'Ein Beitragzur Kenntnis der Seitenbewegung des Tauch-kolbens', *M.T.Z.*, **27**, 5−10 (MIRA Translation No. 39/66) (1966).

[2] Haddad, S. D. and Spata, B., 'Determination of Noise Main Sources of a Two-stroke, Two-cylinder, Diesel Engine to Reduce its Noise', VUNM Res. Inst., Prague (1968).

[3] Fielding, B. J., 'Identification of Mechanical Sources of Noise in a Diesel Engine', *Ph.D. Thesis*, UMIST, 1968.

[4] Haddad, S. D., 'Origins of Noise and Vibration in Vee Form Diesel Engines with Emphasis on Piston Slap', *Ph.D. Thesis*, ISVR, Southampton University, April 1974.

[5] Haddad, S. D. and Pullen, H. L., 'Piston Slap as a Source of Noise and Vibration in Diesel Engines', *Journal of Sound and Vibration*, **34**, 249−260 (1974).

[6] Haddad, S. D., Priede, T. and Pullen, H. L., 'Relation between Combustion and Mechanically-induced Noise in Automotive Diesel Engines', *15th FISITA Congress*, Paris (1974).

[7] Haddad, S. D., 'Liner Deformation due to Piston Slap in Diesel Engines', *4th World Congress on the Theory of Machines and Mechanisms*, Newcastle-upon-Tyne, September (1975).

[8] Haddad, S. D., 'Study of Diesel Engine Noise and Vibration Sources Using Simulation Techniques', *16th FISITA Congress*, Tokyo, 16−21 May (1976).

[9] Haddad, S. D., 'Study of Diesel Engine Noise and Vibration Sources Using Simulation Techniques', *Journal of Automotive Society of Tunku Abdul Rahman College*, Malaysia (1976−77).

[10] Haddad, S. D. and Fortescue, P., 'Simulating Piston Slap by an Analogue Computer', *Journal of Sound and Vibration*, **52**, 791−793 (1977).

[11] Haddad, S. D., 'Mechanically-induced Noise and Vibration in the Automotive Diesel Engine', *ASME Paper 77-DET-37*, Chicago, 26−30 September (1977).

[12] Haddad, S. D., 'A Diesel Engine with Lower Mechanical Noise', *Inst. of Acoustics Meeting*, Cambridge University, 5−7 April (1978).

[13] Haddad, S. D. and Howard, D. A., 'Analysis of Piston Slap-induced Noise and Assessment of Some Methods of Control in Diesel Engines', *SAE Oral Presentation No. 790275,* Detroit, February (1979).

[14] Haddad, S. D., 'Piston Slap Simulation Rig — a Fundamental Tool to Study Piston Slap-induced Vibration and Noise in Diesel Engines', *Internoise 79 Conference,* Warsaw, Poland, 11—13 September (1979).

[15] Haddad, S. D., 'Study of Piston Slap-induced Noise and Vibration in Diesel Engines', *SERC—UNICEG Symposium 'Research in Internal Combustion Engineering in U.K. Universities and Polytechnics',* King's College, London, April (1980).

[16] Grover, E. C. and Anderton, D., 'Noise and Vibrations in Transmissions', *Second International Power Transmission Conference,* Paper No. 7 (1970).

[17] Opitz, H., 'Noise of Gears', *Philosophical Transactions of The Royal Society,* **263,** 461—480 (1968).

[18] Raff, J. A., 'Some Studies into the Origins of Noise of Small Automotive Petrol Engines', *Ph.D. Thesis,* ISVR, Southampton University, 1972.

[19] Russell, M. F., 'Noise from Diesel Fuel Injection Equipment', *FISITA,* Paper 1/7 (1972).

[20] Fawcett, J. N., 'Maintaining Contact Brings Rewards', *Engineering,* 741—743, Sept. (1975).

9

Design options for low noise in I.C. engines

D. Anderton, University of Southampton

1. INTRODUCTION

For some years now it has been appreciated that the noise produced by I.C. engines is intimately linked to the total design concept. Thus decisions taken at the back of an envelope stage of design can have as much control over the final engine noise level as the more complex design details carefully worked out by design groups subsequently. In general, the further the design process has progressed (and almost to the inverse of the exponential law) the smalller is the chance of achieving substantial changes in the final engine noise level. Thus in design for noise, as in design for performance and economy, a continual process of design idea and consequence and trial and result is followed.

2. OVERALL STRATEGIES

Figure 1 shows the design options which affect bare engine noise radiation from radiation from I.C. engines (inlet, exhaust, fan noise, etc. are not considered). There are three stages which can be considered. First — the most rare case — complete design freedom. Secondly, the most common starting point, development of an existing engine design. Thirdly, the situation usually faced, by the customer, installation of the existing engine.

3. COMBUSTION AND INERTIA

The chosen combustion characteristics of an engine determine the primary exciting force applied to the structure and control the design of the engine load carrying structure. The actual structural loads are determined by:

(a) Form and magnitude of the gas force.
(b) Form and magnitude of the inertia forces.
(c) Movement of components in clearances and component deflections.

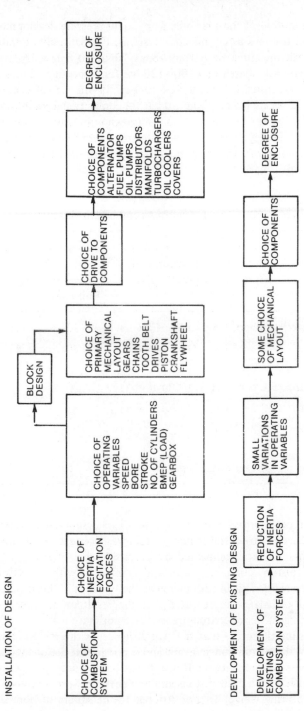

Fig. 1 – Design options for low I.C. engine noise.

The form and magnitude of the gas force as cylinder pressure development are illustrated in Fig. 2 for various combustion systems. For naturally aspirated diesel engines, the peak pressures vary from about 70 to 80 bar, with levels around 100 bar for two-cycle diesels and 100–120 bar for turbocharged diesels. The light load values are usually around 40–50 bar. In the petrol engines, however, the peak pressures are lower, about 40 bar for conventional combustion chambers and about 80 bar for fast burn, high compression pressure chambers. At light load conditions, however, very low peak pressures are found because of intake throttling. The range of rates of pressure rise present in these forms of combustion are shown in Table 1.

Table 1. Rates of pressure rise for various combustion systems

Combustion system	Average maximum rate of pressure rise estimated from measured noise		Peak cylinder pressure – bar	
			Full load	No load
N.A. vee-form four-stroke	8.0 bar/°CA	(117 psi/°CA)	70	43
N.A. in-line four-stroke	7.2 bar/°CA	(105 psi/°CA)	70	43
In-line and vee turbocharged four-stroke	5.0 bar/°CA	(72 psi/°CA)	120	55
I.D.I. four-stroke (N.A.)	3.2 bar/°CA	(47 psi/°CA)	65	60
Roots blown two-stroke	5.0 bar/°CA	(72 psi/°CA)	100	55
Four-stroke petrol engines	2.2 to 4.0 bar/°CA	(32 to 58 psi/°CA)	55	5

The difference in peak pressure of 3 to 1 between combustion systems means a 9 to 1 difference in structural loading for a given bore size at full load. This additional load has to be accompanied by a considerable increase in the basic strength of the piston, connecting rod and crankshaft with a resulting increase in component weight. This in turn increases the loadings in the structure due to inertia and out-of-balance forces, and it is the resultant of these two forces at the main bearings which the structure sees as its primary load.

One of the major inertia forces that the structure sees is the reciprocating inertia force caused by the piston gudgeon pin and a portion of the connecting rod. If the design of these components is put into two groups – deep pistons and high–peak pressure loads and shallow pistons and low peak pressure loads – it is found that the reciprocating weight is proportional to the square of bore size

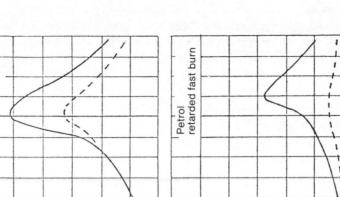

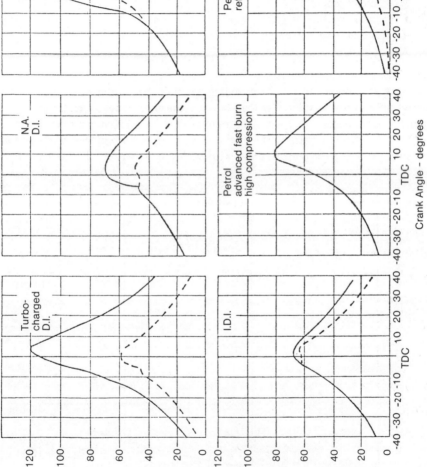

Fig. 2 – Typical cylinder pressure diagrams for various combustion systems.

and that the ratio of maximum inertia force to maximum combustion force is given simply as

$$\frac{\text{Inertia force}}{\text{Combustion force}} = K \frac{N^2 R}{P} \left(1 + \frac{R}{L}\right)$$

where

N = engine speed in rev/min
R = crank radius
P = peak pressure at T.D.C.
L = connecting rod length
K = constant.

and is independent of bore size.

Thus the key factors in determining whether the combustion pulse or the inertia pulses provide the larger load become the engine speed, crank radius and cylinder peak pressure at T.D.C. The relationship for a 150 mm stroke is shown in Fig. 3. The figure illustrates the overwhelming importance of the cylinder peak pressure. At low peak pressures and with such a large stroke, the inertia forces

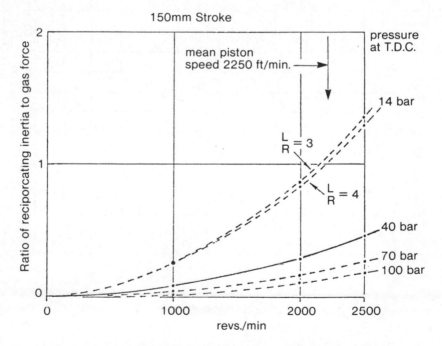

Fig. 3 – Ratio of maximum inertia and gas forces due to reciprocating unbalance as a function of engine speed for 150 mm stroke.

very rapidly become predominant as engine speed increases. The limiting piston speed is marked on the figure. For peak pressures which occur within the range of diesel combustion, the magnitudes of the inertia forces are always less than one-quarter of those due to combustion. Only at very low peak pressures associated with petrol combustion would inertia forces become the dominant concern. Figure 4 shows a similar calculation for a 75 mm stroke engine with very light (i.e. petrol engine) pistons. Again, even up to 6000 rev/min for diesel type peak pressures, the inertia forces are well below those due to combustion. However, at typical light load petrol engine peak pressure values, the inertia forces become several times those due to combustion. The L/R ratio has only a small effect on the magnitude of the forces. Of course the inertia load is always applied in the four-stroke engine to its full extent on the induction stroke but its absolute magnitude does not exceed that due to the full load combustion pressure (except perhaps on over-run), particularly as compression ratios are increased to improve economy.

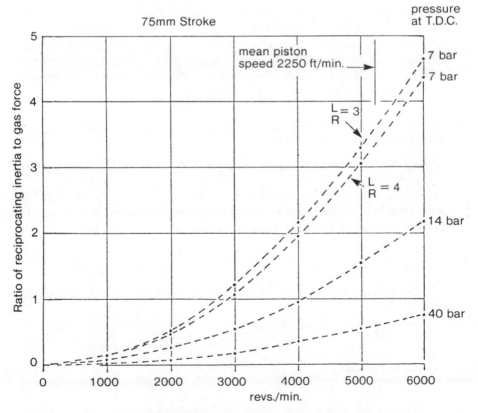

Fig. 4 – Ratio of maximum inertia and gas forces due to reciprocating unbalance as a function of engine speed for 75 mm stroke.

The form of this combined combustion and inertia loading at the main bearing is illustrated in Figs. 5, 6 and 7. Figure 5 shows the applied force variation for the gas loading shown in Fig. 1 for a turbocharged diesel at full load and with an inertia force to gas force (I/G) ratio of 0.17 (i.e. low engine speed) and also for a 'normal' petrol engine with the same I/G ratio. The total predominance of combustion in both cases is very clear. It is also clear that the combination of combustion and inertia can provide high rates of change of force level ($d^2F/d\theta^2$), which contribute to high frequency noise. Figure 6 shows a similar calculation for the high speed case of D.I. combustion (4000 rev/min) in which the I/G ratio is 0.86 and the same petrol diagram in which the I/G ratio is 1.08. In both cases, the maximum force levels are controlled by inertia and the combustion pulse in combination with the inertia provides high rates of change of force level. In these cases the fundamental excitation frequency is the cycle frequency. If the combustion pulse is removed (closed throttle on the petrol engine) then the force diagram becomes symmetrical (Fig. 7) and the fundamental excitation frequency is the rotation frequency.

The effect that these changes have on noise is illustrated by the noise level of two engines at full and no load (Fig. 8). In the case of the diesel engine, where the change in peak pressure is small, the engine noise does not change substantially with load. However, for the gasoline engine there is a substantial difference between the full load and no load noise at the same speed, illustrating the importance of peak cylinder pressures on noise.

4. PISTON SLAP AND INERTIA

As well as the overall effect of inertia loads there is the associated excitation of piston slap. Noise radiation due to both these sources can be produced by either the kinetic energy of impact (K.E.) or the direct forcing of the structure by a load (side load in piston slap) — a sort of potential enenergy term (P.E.). The radiated noise intensity (I) for these sources can be approximated as [1]:

For Inertia Excitation

$$I_{\text{K.E.}} \propto M.N^2.\delta^{\frac{4}{3}}$$

$$I_{\text{P.E.}} \propto M^2.N^6.\delta^{\frac{2}{3}}$$

For Piston Slap

$$I_{\text{K.E.}} \propto M^{\frac{1}{3}}.N^{\frac{2}{3}}.\delta^{\frac{4}{3}}$$

$$I_{\text{P.E.}} \propto M^{\frac{2}{3}}.N^{\frac{10}{3}}.\delta^{\frac{2}{3}}$$

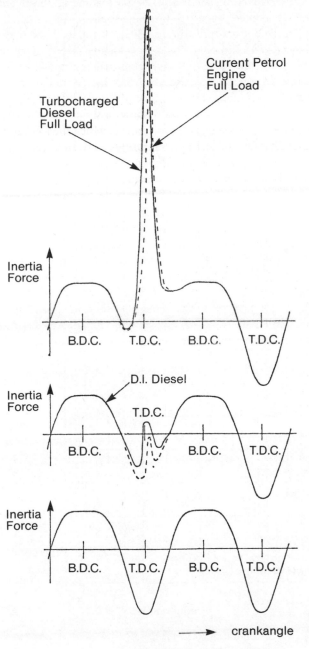

Current Petrol
Engine
Full Load

Turbocharged
Diesel
Full Load

Inertia
Force

B.D.C.   T.D.C.   B.D.C.   T.D.C.

Fig. 5 – Main bearing loads
for turbocharged diesel and
petrol engines at low speed.
2000 rev/min (ratio of
inertia to gas loads 0.17).

D.I. Diesel

Inertia
Force

T.D.C.

B.D.C.   T.D.C.   B.D.C.   T.D.C.

Fig. 6 – Main bearing loads
for D.I. diesel and petrol
engines at high speed. 4000
rev/min (ratio of inertia to
gross loads 0.86 and 1.08).

Inertia
Force

B.D.C.   T.D.C.   B.D.C.   T.D.C.

crankangle

Fig. 7 – Main bearing loads
for petrol engine at high
speed and closed throttle.

COMBUSTION AND INERTIA

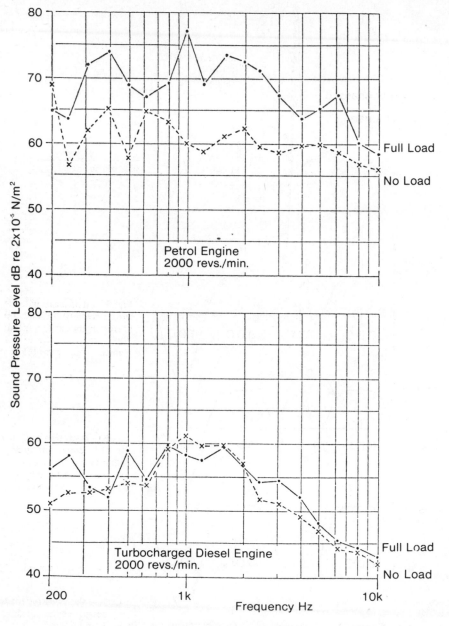

Fig. 8 – Effect of load on engine noise.

where

M = reciprocating mass
N = engine speed
δ = clearance.

The importance of the inertia (P.E.) term at high engine speeds can be readily seen, and the relationship with combustion is illustrated in Fig. 9.

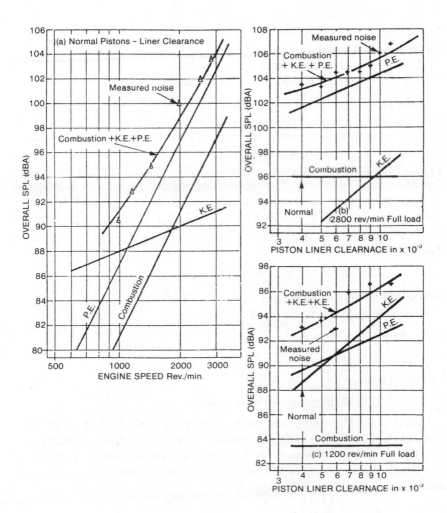

Fig. 9 – Relation between inertia excitation combustion and noise.

5. COMBUSTION NOISE MODEL: EFFECT OF SYSTEM ON RESULTANT NOISE

A simple, linear combustion noise model for I.C. engines can be devised [2] from which the noise intensity is given by

$$I \propto P_c . N^n . B^4 . S_{rad} \, C_{A.S.O.}$$

where

P_c	= pressure level of reduced and simplified cylinder pressure spectrum at $f/N = 1.0$
f	= frequency
N	= engine speed
n	= combustion index
B	= engine bore
S_{rad}	= engine surface area
$C_{A.S.O.}$	= overall combined acoustic and structure response of the engine.

The product $P_c . N^n . B^4$ defines the force applied to the engine structure by the combustion process, whereas the product S_{rad} defines the ability of the structure to alternate and radiate that force. Figure 10 illustrates the relationship between combustion system, engine bore and radiated noise at constant speed. The radiated noise is dependent on B for all combustion systems except on gasoline engines and the level of noise is seen to depend greatly on the choice of combustion system. The average measured variation in the absolute level of combustion excitation, P_c (expressed as $20 \log_{10}(P_c/P_{ref})$ and n, can be expressed as

Two-stroke Diesel

$$CP_R \, (1.0) \, = \, 160.5 \text{ dB} \, ; \, n \, = \, 4.15$$

Four-stroke, Normally Aspirated Diesel – Indirect Injection

$$CP_R \, (1.0) \, = \, 145.0 \, ; \, n \, = \, 4.45$$

Turbocharged Diesel Engines (full load)

$$CP_R \, (1.0) \, = \, 144.0 \, ; \, n \, = \, 5.10$$

Gasoline Engines

$$CP_R \, (1.0) \, = \, 133.0 \, ; \, n \, = \, 6.30$$

Thus for a given bore and speed, a comparison of the basic combustion noise levels can be made (in detail if cylinder pressure spectra are available).

The empirical relationship with bore of B^5 can be used to estimate the character of the engine response factor $C_{A.S.O.}$. Since for most engines, $S_{rad} \propto B^2$ rather than B^3, then we can imply that

$$C_{A.S.O.} \propto \frac{1}{B}$$

in general terms for engines of conventional constructuion.

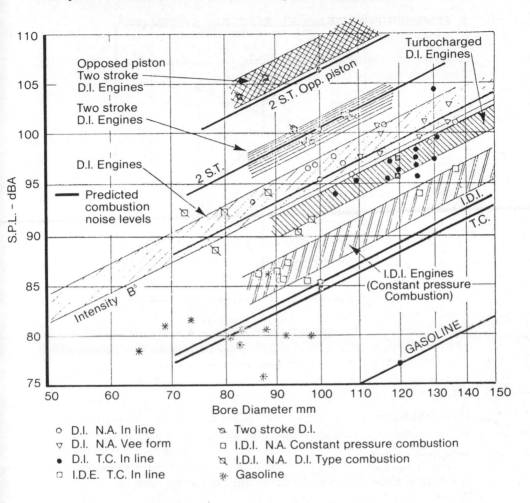

Fig. 10 – Effect of combustion system and bore size on engine noise.

6. OPERATING PARAMETERS

Three of the key design parameters affecting engine noise are combustion system, speed and bore. The relationship for engine noise as a function of bore is illustrated in Fig. 10. It can be seen that the measured and predicted effect of a combustion system on noise is to change it, for a given engine speed, by up to 20 dBA. This is illustrated further in Fig. 11 where the full load noise versus engine speed noise levels of four 100 mm bore engines are shown. At 2000 rev/min there is a 16 dBA noise difference. The strokes of the engines shown are 110 mm, 108 mm, 86 mm and 76 mm. The rated speed is determined by the

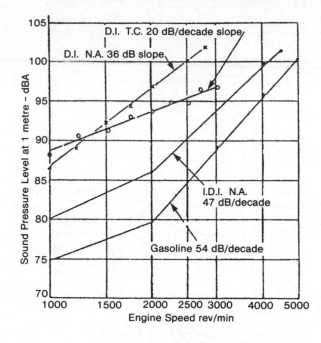

Fig. 11 − Effect of engine speed on noise of engines of 100 mm bore and different combustion systems.

stroke which in turn, and if the engine is run to its limiting piston speed, determines whether combustion excitation or inertia excitation is predominant. At rated speed, the maximum engine noise is about the same for all these engines. It is clear that as engine stroke is increased the rate of increase of engine noise with speed is also reduced from a fifth power to a third power relation.

7. SPECIFIC WEIGHT OF PRESENT ENGINE DESIGNS

As far as engine weight is concerned, the relationship between emgine specific weight (kg/litre) and cylinder volume (bore) is shown in Fig. 12 for six-cylinder diesel engines and in Fig. 13 for four-cylinder diesel engines. For any given cylinder capacity there is a huge variation in engine weight. The six-cylinder engine data are shown again in Fig. 14, but for radiated noise at rated conditions. Again, for any given noise level there is a considerable variation in specific weight. There is a tendency for turbocharged engines to lie on the left of the figure. The noise levels of some experimental engines with modified structures are also indicated. At present it is clear that there is no direct relationship between engine specific weight and engine noise.

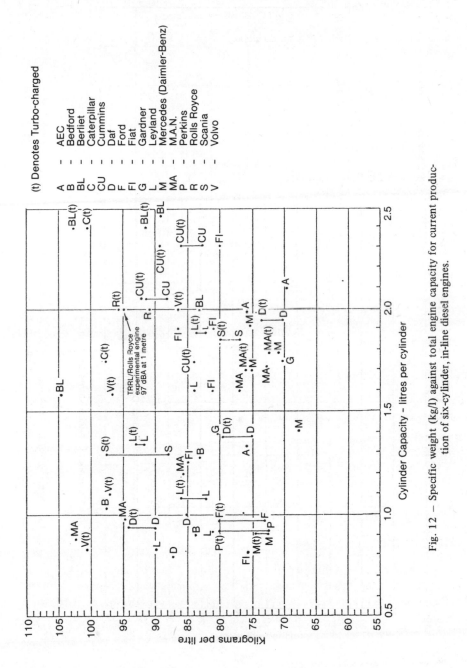

Fig. 12 – Specific weight (kg/l) against total engine capacity for current production of six-cylinder, in-line diesel engines.

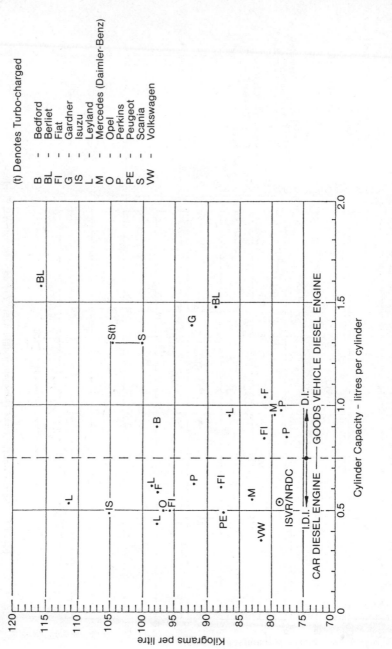

Fig. 13 — Specific weight (kg/l) against total engine capacity for current production of four-cylinder diesel engines.

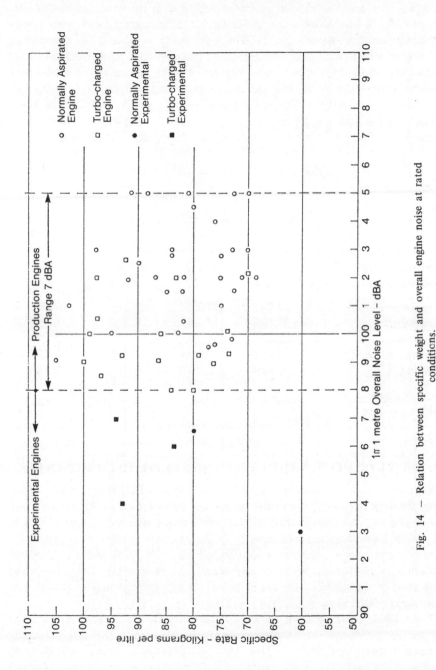

Fig. 14 — Relation between specific weight and overall engine noise at rated conditions.

The percentages of total engine weight due to pistons, connecting rod, crankshaft, flywheel and crankcase/block for three larger diesel engines (200–350 bhp) and one petrol engine (100 bhp) are shown in Table 2. The distribution of engine weight according to component is surprisingly similar between large diesel and small, high speed petrol engines as is the percentage of total engine weight contributed by the rotating and reciprocating components (roughly half the total engine weight). The reduction in specific weight between the heavy diesel and the high speed petrol engine is between 35 and 40%, but at the same time the specific power increases from about 0.25 bhp/kg to 0.7 bhp/kg.

Table 2. Component weights of engines

Bore and type	130 mm bore six-cylinder, in-line diesels			80 mm bore four-cylinder petrol
Specific weight	91.4	86.33	83.78	54.5
Specific power	0.26	0.32	0.26	0.71
Crankcase/block	29.7%	31.2%	28.8%	27.0%
Crankshaft	10.16%	12.8%	8.9%	10.27%
Flywheel	5.4%	3.6%	8.2%	4.0%
Connecting rod [1]	0.33%	0.57%	0.42%	0.52%
Piston [1]	0.39%	0.46%	0.45%	0.35%
Sum of above as % of total weight	52%	54%	52%	48%

8. CHOICE OF MECHANICAL LAYOUT AND DRIVE TO COMPONENTS

In the late 1950s, research was carried out on the possibilities for demonstrating that the diesel engine noise can also be reduced by an entirely different approach to engine structure design [3]. In this work which was carried out at C.A.V. Research Laboratories, two alternative principles were demonstrated, namely a space frame engine structure with damped outer surfaces and a stiff engine structure of lightweight material (magnesium). These investigations proved that the noise of the diesel engine can be halved (in subjective terms) by attention to structural details alone.

In 1964, resulting from an S.R.C. grant, a production feasible low noise engine was produced which preceded similar thoughts from other internationally known research establishments (Fig. 15). This design was initially applied to an engine in the passenger car size range and it was followed by experimental engines in the medium truck range from 100 to 160 hp of in-line and V8 forms.

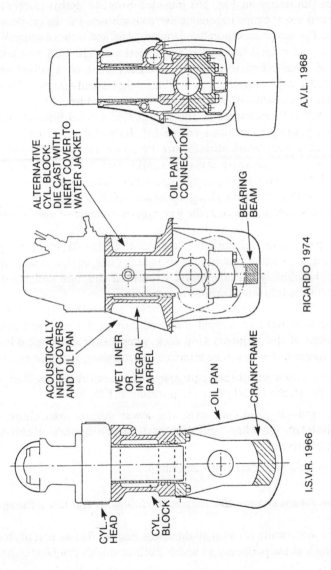

ALTERNATIVE
CYL. BLOCK:
DIE CAST WITH
INERT COVER TO
WATER JACKET

OIL PAN
CONNECTION

BEARING
BEAM

A.V.L. 1968

ACOUSTICALLY
INERT COVERS
AND OIL PAN

WET LINER
OR
INTEGRAL
BARREL

OIL PAN

CRANKFRAME

RICARDO 1974

CYL.
HEAD

CYL.
BLOCK

I.S.V.R. 1966

Fig. 15 – Crankframe engine development.

In 1972, I.S.V.R. designed and built for the T.R.R.L. Quiet Heavy Vehicle (Q.H.V.) Project two in-line six-turbocharged engines of 210 and 350 hp. The novel features (illustrated in Fig. 16) included bedplate design integrally connecting the main bearing caps to provide extreme stiffness for the crankcase area and rear gears. The design incorporated damped panel and isolated sump. With no weight increase, an overall test bed noise reduction of 10 dBA was obtained. I.S.V.R. was also associated with the lightweight (about 60 kg/litre) bedplate design of an 8 litre in-line diesel engine with a foreign manufacturer [4], resulting in an extremely quiet engine of about 93 dBA at 1 m distance.

In 1977, N.R.D.C. sponsored an investigation of a novel lightweight engine structure for a passenger car diesel engine [5]. Its essential feature is a simple lightweight casting capable of manufacture by conventional techniques, and is illustrated in Fig. 17. Crankcase walls are cantilevered from a rigid lower deck with designed low natural frequency which are largely separated from the crankshaft supporting structure giving vibration isolation (Fig. 18). To reduce weight, noise and construction cost, the water jacket consists of separate damped panels.

These proven design schemes for low noise engine structure designs are illustrated in Fig. 19. Central conclusions are hard to draw, but based on economic, production and durability considerations, a universal structure design for heavy duty diesels with the following features is suggested.

(1) An engine with walls of normal thickness cast as flat as practicable with raised flanges at the periphery (top deck, sump flange and block side edges) to which damped shields can be attached (by screwing, gluing or riveting).

(2) Rear gear location with heavy cast gear cover to which all ancillary equipment can be attached (fuel pump, compressor, P.T.O., etc.)

(3) Crankcase split at crankshaft axis, the lower portion comprising a stiff integral bedplate providing rigid support to which the side shields and oil sump can be attached.

(4) Damped or isolated rocker cover and sump.

For lightweight diesels and gasoline engines the following features are suggested.

(1) An engine with walls of normal thickness cast as flat as practicable with raised flanges at the periphery to which damped shields can be attached.

(2) Tooth belt or chain drive to ancillary equipment, which should be mounted to sturdy sections of the engine.

(3) Normal crankcase split but specially designed main bearing bulkheads and crankshaft.

(4) Damped or isolated rocker, sump, front cover and manifolds.

1. AREA FOR FIXING
   ANCILLARY EQUIPMENT

2. STIFF BED PLATE

3. RAISED CAST FLANGE
   FOR COVER ATTACHMENT

4. DAMPED SIDE SHIELD

5. ISOLATED OIL PAN

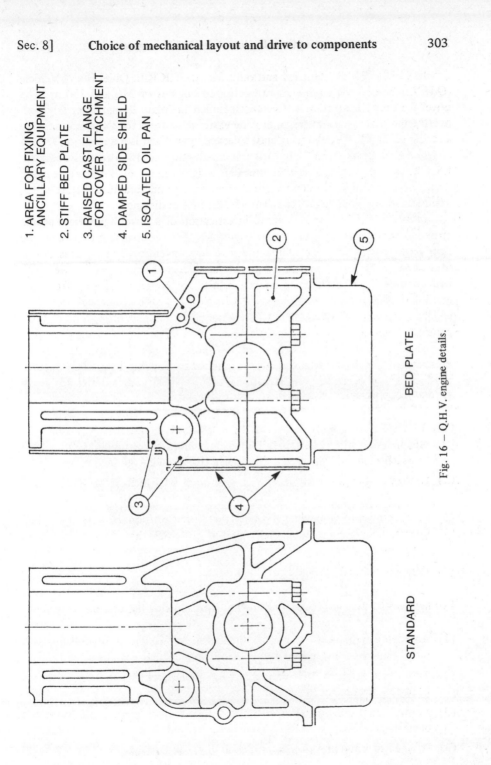

BED PLATE

STANDARD

Fig. 16 – Q.H.V. engine details.

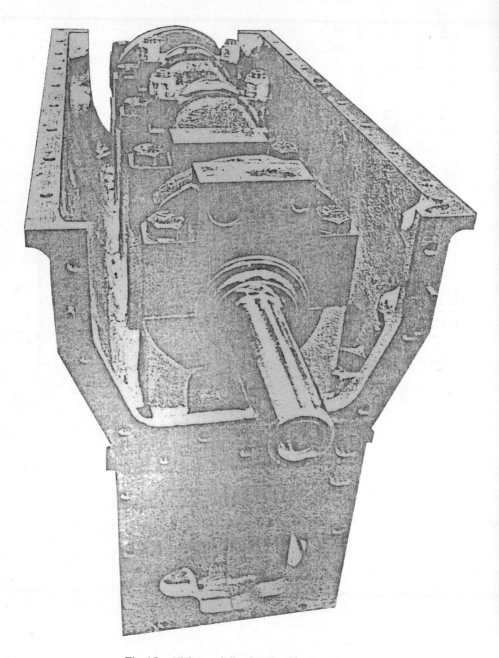

Fig. 17 – High speed diesel engine block structure.

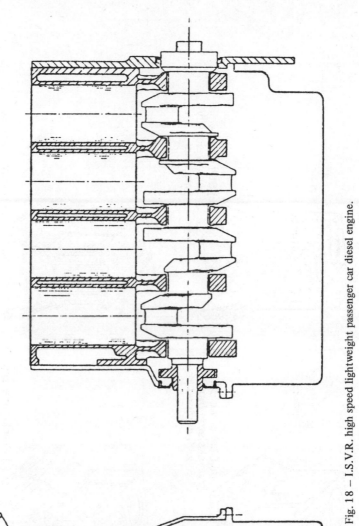

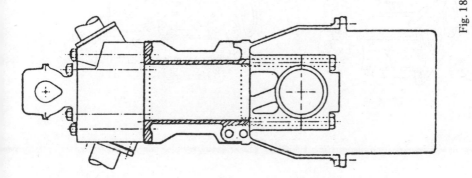

Fig. 18 – I.S.V.R. high speed lightweight passenger car diesel engine.

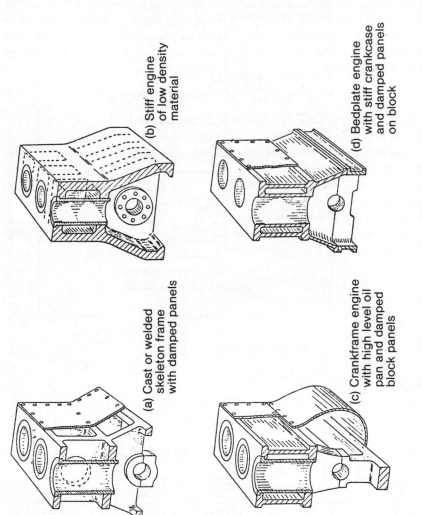

(a) Cast or welded skeleton frame with damped panels

(b) Stiff engine of low density material

(c) Crankframe engine with high level oil pan and damped block panels

(d) Bedplate engine with stiff crankcase and damped panels on block

Fig. 19 – Proven schemes for low noise engine design.

9. CHOICE AND DESIGN OF COMPONENTS

Where there is a choice of component suppliers, such as for alternators, fuel pumps, turbochargers, etc., some differences in noise radiation charactieristics may be used. Drives to alternators and fuel pumps are critical. Gear driven alternators are noisy and any appreciable backlash in fuel pump drives should be avoided. Other components, such as manifolds, covers, etc., can be treated in a number of ways. The noise radiated by a surface of area S_{rad} and area average mean square velocity

$$\text{S.P.L.} \propto 10 \log_{10} S_{rad} + 10 \log_{10} \sigma_{rad} + 20 \log_{10} \langle U \rangle$$

where σ_{rad} is the acoustic radiation ratio of the surface.
S.P.L. = Sound Pressure Level.

The noise radiated is most affected by surface velocity and thus any reduction in this (by damping, stiffening or detuning) will be of benefit. Reduced component surface area should always be a design aim. Significant reductions in the radiation ratio σ_{rad} are only possible by using thin sheet construction in place of castings.

A simple general philosophy is illustrated in Fig. 20. Small steel pressings

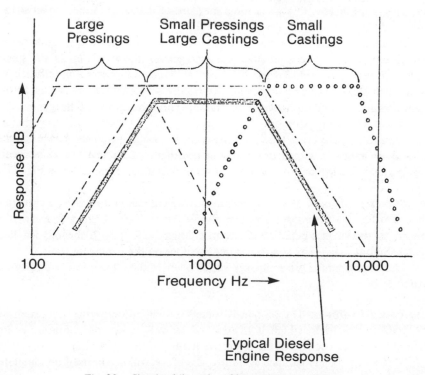

Fig. 20 – Simple philosophy of low noise cover design.

and large iron or aluminium castings respond well at engine excitation frequencies in the middle range and should therefore be avoided. For large covers it is more advantageous to aim at a design using steel pressings where the natural frequencies will be low and because of the low stiffness damping techniques can be readily applied. For small covers, castings could be more suitable because the greater stiffness would result in high natural frequencies above the critical range.

Basic low noise cover designs are described in the following sections.

9.1 Oil Pan

(a) *Isolation:* Requirement here is an effective isolation without using an expensive chemical rubber bonding process. Adequate oil sealing and fail-safe properties are mandatory.

Figure 21 illustrates a method for a cast oil pan. A cast aluminium ring of cross-section as shown traps the oil pan flange in a U section rubber extrusion. Because no washer is needed between the clamping ring and the crankcase flange, and the damping ring itself is of a heavy section, no local distortion occurs at the fixing screws. Clamping pressure on the oil pan flange is therefore uniform around the periphery and oil leaks are obviated.

This form of assembly gives some measure of damping to the oil pan structure in addition to some measure of isolation.

On some present production engines the oil pan flange on the crankcase is insufficiently wide to accommodate the arrangement shown in Fig. 21 and some modification is required. Figure 21 shows examples of possible modifications using thick steel stampings to extend the crankcase skirt width. However, an oil sealing washer is required between the stamping and the crankcase flange.

Where the sump structure is attached to the flywheel and transmission housing and is therefore a load carrying member, isolation techniques are more difficult to apply. In some cases it may be possible to use a heavy brace between the flywheel housing and the isolated oil sump fixing flange as shown in Fig. 21.

(b) *Increased stiffness:* Where an increase of weight can be tolerated, a cast-iron oil pan offers an advantage. The pan should be heavily ribbed and internally braced as shown in Fig. 21. With such a design, not only is the oil pan less responsive but it also acts as a restraining member on the crankcase walls.

This arrangment is pre-eminently suitable where the oil pan is a load carrying structure.

(c) *Reduced stiffness:* The oil pan is forced into vibration by the movement of the crankcase walls to which it is attached as well as set into vibration at its natural frequency because of the impulsive nature of the excitation.

A non-stiff, highly damped structure, therefore, will not in itself be adequate to control oil pan noise.

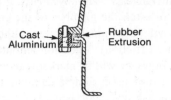

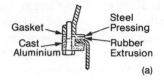

(a)

GENERAL ARRANGEMENT FOR
ISOLATION OF CAST OIL PAN

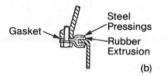

(b)

ISOLATED OIL PAN ARRANGEMENT
WHERE PAN FLANGE IS NARROW

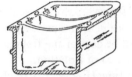

HEAVY CAST IRON OIL PAN

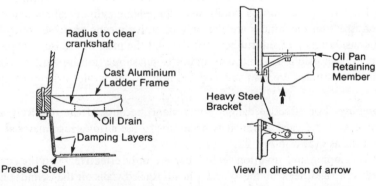

View in direction of arrow

GENERAL ARRANGEMENT FOR
PRESSED STEEL DAMPED OIL PAN
WITH LADDER FRAME CRANKCASE
RESTRAINING MEMBER

FLYWHEEL HOUSING SUPPORT
WITH ISOLATED OIL PAN

Fig. 21 – Low noise oil sump designs.

Figure 21 illustrates a technique using a stiff ladder frame to tie the crank-case walls in conjunction with a pressed steel damped oil pan. The ladder frame, being laterally stiff, goes some way to restrain crankcase wall movements. If only normal sized oil pan fixing screws are used to fasten the member to the crank-case it is likely that crankcase wall movement will not be much affected (although some damping will be introduced by interfacial friction). Vibration amplitudes at the oil pan fixing face on the ladder frame, however, will be markedly reduced, regardless of the size and number of fixing screws used. Simple damping tech-niques can be applied to the pan surface, example pads spot welded or glued to local areas as shown. For optimum damping it is essential that the oil pan surfaces should consist of large, flat areas — the applied damping members should most certainly be flat and should not extend to the corners of the pressing.

9.2 Valve cover

(a) *Isolation:* In Fig. 22, a conventional design of flange mounted cast aluminium valve cover is illustrated which has been split (or cast as two separate pieces) and chemically bonded. The plane of the bond should be as near the fixing flange as practicable in order to ensure the greatest possible area of isolation.

Where the design is of the centre bolt fixing type (Fig. 22), a thick rubber washer can be used to isolate the cover flange from the cylinder head surface in conjunction with rubber grommeted fixing screws.

Where the bank-to-bank mode of vibration is a problem on vee form engines, that is the valve cover moves bodily with the engine cylinder block and head, vibration isolation techniques are the only effective means of noise control. By careful consideration of the rubber stiffness and the mass of the isolated part of the cover it is possible to achieve an opposite phase relationship with the bank. Vibration amplitudes are also reduced because of the induced damping at the bend.

(b) *Damping:* For effective damping treatment, because of the relatively small physical size of the valve cover, it is necessary to use a thin gauge material with flat surfaces, as shown in Fig. 22.

Both damping and isolation techniques are only effective when the cover is attached to a rigid deck of the cylinder head. Raised walls on the cylinder head, which are often introduced to reduce the valve cover depth and also to provide a barrier to effect more efficient oil sealing, are to be avoided.

9.3 Cylinder Head Shield

Where cover isolation and damping techniques are not practical and where a large area of head surface remains exposed, total shielding of the upper cylinder head (including possibly the injectors) is an effective means of noise control. The shield, which is constructed of lightweight damped material, is fixed to the valve cover via a rubber grommet as shown in Fig. 22. Rubber piping on the rim offers damping, isolation and air sealing. Oil tight sealing is not required.

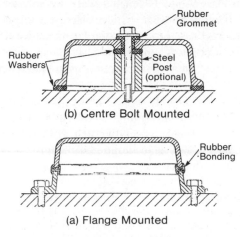

(b) Centre Bolt Mounted

(a) Flange Mounted

ISOLATED VALVE COVER DESIGNS

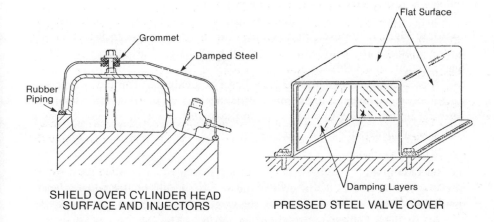

SHIELD OVER CYLINDER HEAD
SURFACE AND INJECTORS

PRESSED STEEL VALVE COVER

Fig. 22 — Low noise valve cover designs.

9.4 Gear Case

The front gear case can be a formidable noise control problem. Often it is directly loaded by ancillary equipment such as water pump, cooling fan, injection pump drive bearing, air compressor and auxiliary power take-off.

A simple pressed steel case can be treated by methods previously outlined for the oil pan and valve cover. A simple cast cover can in some cases be isolated using the flange and U section rubber extrusion technique described for the oil pan.

In many cases it is only possible to treat part of the surface as shown in Fig. 23. Here the treated section can be removed and bonded in position as in Fig. 23, or a separate piece can be bonded over the area as in Fig. 23. This provides effective damping in shear as well as in vibration isolation.

Figure 23 is a stiff peripheral cast flange with a separate casting consisting of the whole of the cover front surface bonded into it.

Where weight is not a limiting factor, noise control can be effected by using a heavy cast-iron cover.

9.5 Exhaust Manifold

Ideally, the designer should ensure that the surface area of this member is as small as practicable. Round sections are to be preferred to square ones to ensure minimum surface area for maximum flow volumes.

Sheet steel covers can be simply made to enclose the exhaust manifold as shown in Fig. 24. However, to minimize noise built up in the enclosure and to ensure optimum attenuation of the cover, it is necessary to line the cavity with a heat resistant, sound absorbent mineral wool.

A problem associated with enclosed cast-iron exhaust manifolds, especially on turbocharged engines, is the internal flaking that occurs at elevated temperatures. Therefore it could be necessary to use a steel manifold or to cast from a high quality nickel iron alloy.

9.6 Inlet Manifold

Shielding techniques as described for the exhaust manifold also apply to inlet manifolds.

Alternatively, vibration isolation is possible. The problem is somewhat simpler on a vee-form engine where a simple technique using short rubber pipes secured by hose clips can be used as shown in Fig. 24. In the case of turbocharged engines the effect of inlet air pressure must be taken into consideration.

On in-line engines, some degree of isolation and damping can be obtained using composite washers as shown in Fig. 24. These washers consist of a steel entral layer bonded between two thick rubber layers. Thus a thick isolating washer is obtained which cannot be squeezed out when the joint is lightened because of the restraining action of the central steel shim.

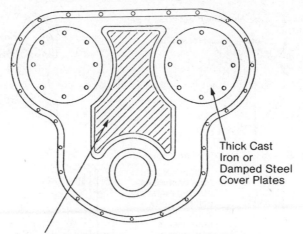

Thick Cast
Iron or
Damped Steel
Cover Plates

Largest possible free area of surface highly
damped for steel pressings or removed and
replaced with rubber bonding for castings

FRONT TIMING GEAR COVER DESIGN

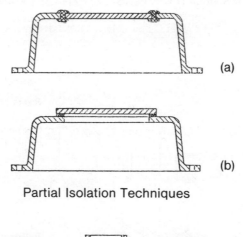

(a)

(b)

Partial Isolation Techniques

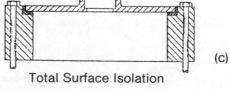

(c)

Total Surface Isolation

GEAR CASE DESIGNS

Fig. 23 – Low noise front timing gear cover designs.

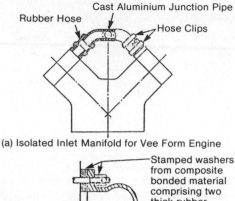

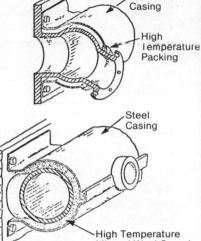

(a) Isolated Inlet Manifold for Vee Form Engine

(b) Isolated Inlet Manifold For In Line Engine

ISOLATED INLET MANIFOLD DESIGNS   EXHAUST MANIFOLD SHIELDING

Fig. 24 – Low noise manifold designs.

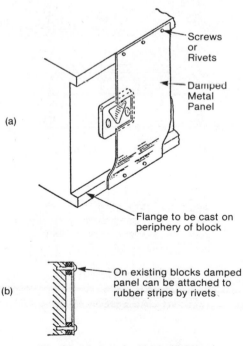

Fig. 25 – Low noise block shields.

9.7 Block Shields

Block shields are most effective when the engine structure has been designed to accommodate them. The ideal block structure, as illustrated in Fig. 25, has flat walls with a cast-in flange around the periphery for cover attachment. All engine accessories and ancillary equipment are mounted after the panel is attached; thus the panel is fitted during engine assembly (by adhesives rivets or screws) and is not subsequently removed.

10. DEGREE OF ENCLOSURE

In stationary installations there is considerable scope in the design and construction of an engine house to provide very large noise attenuations. In vehicular applications the degree of noise attenuation is rather more dependent on the type of vehicle and the degree of enclosure afforded in its concept. This is illustrated in Fig. 26. Enclosure is not welcomed by either customer or manfacturer but often is the most cost effective solution for minor infringements of noise targets.

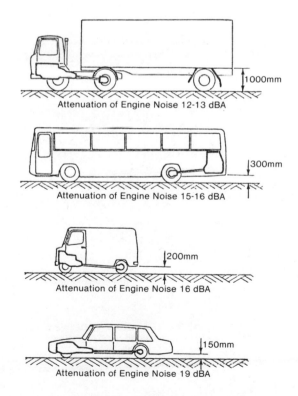

Fig. 26 – Attenuation of different vehicle designs.

REFERENCES

[1] Lalor, N., Grover, E. C. and Priede, T., 'Engine Noise due to Mechanical Impacts at Pistons and Bearings', *SAE Paper 800402*.

[2] Anderton, D., 'Relationship between Combustion System and Engine Noise', *SAE Paper 790270*.

[3] Priede, T., Austen, A. E. W. and Grover, E. C., 'Effect of Engine Structure on Noise of Diesel Engines', *Proc. I. Mech. E., 179*, part 2A, No. 4.

[4] Ochiai, K., and Yokota, K., 'Light Weight, Quiet Automotive D. I. Diesel Engine Oriented Design Method', *SAE Paper 820434*.

[5] Grover, E. C. and Perry, R. D. H., 'An Experimental Passenger Car Diesel Engine', *SAE Paper 790443*.

10

Ensuring the reliability of diesel engine components

D. A. Alcraft, L. Gardner Diesels

1. INTRODUCTION

The two principal reasons for the commercial success of the compression ignition oil engine are its superior efficiency and reliability when compared to competitive prime movers, particularly those available for use in vehicles. To a small extent, efficiency contributes towards reliability because it implies that the engine is consuming less energy on its own destruction and more is being converted into useful work, so continuing improvements in mechanical and thermal efficiency are important. But as manufacturers seek to increase the power available from an engine or to reduce its weight, design changes are introduced which require testing so as to ensure that the overall reliability of the product is not impaired and is preferably improved. This chapter is concerned with the principles and techniques employed in acquiring assurance of satisfactory reliability.

Any machine may be so designed and built that it will have an almost indefinite life, given proper maintenance, but this philosophy inevitably carries a penalty of cost and weight which severely limits the applications available. The vehicular diesel engine must have minimal weight so as to maximize the pay load which may be carried and must be competitive in price, while it may also be expected to last for perhaps ten years or for up to half a million miles, often with less than perfect maintenance. Even a vehicle which is driven in shifts and covering 1000 miles per day will require two years to reach half a million miles. In contrast, a manufacturer within a commercial environment must acquire confidence in his new product in as short a time as possible and the timescale for the final test needs to be weeks rather than years. Development engineers therefore depend upon specially devised tests combined with previous experience from a wide background in order to make confident and reliable assessments of a component's performance. Thus a theoretical understanding of the operation and loading pattern of each component is combined with experience to guide the search for small defects which may be indicative of serious problems in the longer term. The developed engine is then proved to be reliable by an endurance test which exacerbates the loading and stress of operation found in service without destroying the evidence produced. It is always possible to wreck a machine by misusing it and this is not the objective.

2. EARLY DECISIONS

Obviously the engine designer has a very great influence upon the ease or difficulty of achieving acceptable reliability and has to make various policy decisions which in turn determine the detail design of the engine. As an example, a choice must be made as to whether the engine should have separate cylinder liners in direct contact with the coolant (wet liners) or a cylinder block which is itself sleeved with a thin cylinder to provide a suitable running surface (dry liners). The remaining possibility of a cylinder block without any liner is usually eliminated because of the expense of renewing the whole block in service. It may be considered that superior heat transfer will be achieved with the wet liner, but this is susceptible to the problem of waterside attack (Fig. 1) which could result in a disastrous engine failure if the liner were not renewed in time. This problem may be totally eliminated by the use of dry liners, although there are then other potential difficulties to be avoided such as liner pull down. Thus the choice made will be determined by the knowledge of the designer as to which potential problems his company has the experience and facilities to resolve.

Once a component has been designed it is necessary to obtain prototype samples and this is often not straightforward. A particular difficulty may be that the samples have not been made by the same process as is intended for production, usually on the grounds of tooling cost. Such components are often adequate for initial tests but it is essential that as soon as possible and certainly for the final tests prior to full production release, components produced by the proposed production techniques and tooling should be used. Manufacturing techniques can have a major influence upon the performance of a component.

If only one or two components are particularly different in a new design it may be considered that it would be quicker and more convenient to use a rig rather than an engine for testing them. Great care is required, however, to ensure that the loading applied by the rig is indeed representative of that which will occur in the engine. Each component only has value in its relationship with other components as a part of the whole and it is frequently the relationship to adjacent components which causes problems rather than the mechanical strength of the component itself. Scuffing, which can occur on piston rings, pistons and other parts as well as on valve stems (Fig. 2) is an example of this type of problem. In this instance, lubrication of the valve stem in the valve guide has failed and minute asperities of the two surfaces have welded together through frictional heat. Relative motion has torn the welds apart and smeared them over to provide

Fig. 1 – A wet cylinder liner which has suffered waterside attack. The cause is erosion rather than corrosion and arises from vibrations of the liner.

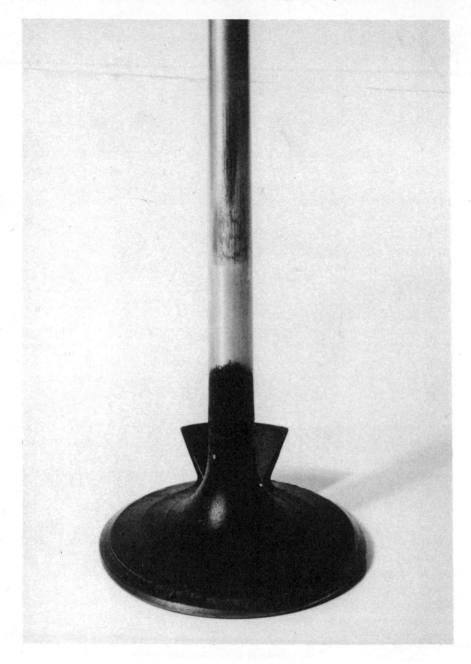

Fig. 2 — A valve stem which has scuffed.

the characteristic scuffed appearance. It is unlikely that this phenomenon could be accurately reproduced on valve gear being tested in a simple rig because the breakdown of lubrication is a function of the transient temperatures to which the valve stem is subjected as well as of the mechanical loadings and supply of lubricant. A very sophisticated rig may achieve reasonable simulation of engine conditions but the cost would only be justified for a major research investigation. When the engine is probably being run for other reasons it is preferable by far to test alternative designs and metallurgies in the engine. Then there will be little doubt that the valve (or other component) is being subjected to the same thermal and mechanical loading which it will have to withstand in service. Rigs have often been used, however, when a major development step has been required on a specified component, in order to allow progress with uprating of an engine. Redesigned fuel injection equipment would be extensively rig tested, for example, but the greater part of a fuel injection system is not subjected to extreme thermal loading and the mechanical loading can be accurately reproduced, so that rig results in this instance are valid.

3. PROTOTYPE ENGINES

The majority of component development is therefore conducted in engines which progress through the stages of development. In the beginning, the major question is whether all the parts will fit together and whether the engine can achieve the desired power output. If major components such as the crankshaft or pistons are of new design it often results in some fairly basic difficulties, particularly if specific loadings are being increased. Provided that adquate care has been taken in the construction of the engine it may reasonably be expected to run, but large numbers of prototypes have failed either before or soon after reaching full power. A frequent cause is either a piston or a bearing failure, dependent upon the design, and in both cases it is essential that the engine be stopped immediately. The power available from the undamaged parts of an engine can very rapidly cause the complete destruction of the component which is in trouble. with the result that all evidence as to where and how the problem began is lost. The remains of a piston shown in Fig. 3 are totally inadequate to allow any sensible judgement as to which part of the piston requires modification. Automatic control of the test bed installation may prevent such disasters, if thermocouples or pressure signals can provide a warning of impending trouble and be used to trigger the shutdown of the engine at the initiation of failure, but for piston seizures or for unexpected events there is no substitute for good reactions from the technician running the engine.

Provided that a seizing engine is stopped quickly, the damage will remain localized and it will then be possible to decide what action must be taken to avoid any repetition. Often other components in the engine will provide a good indication of the mechanism of failure by showing it at earlier stages, such as a

Fig. 3 – Allowing an engine to continue running long after initiation of the piston seizure has caused such extensive damage that it is not possible to discover how or where the problem began. An expensive test is wasted.

crack which has not yet developed into the break which caused the seizure. This would certainly assist in resolving a complicated failure such as that shown in Fig. 4 and which apparently requires corrective action in more than one area. Care must be exercised, however, to ensure that the fault actually lies with the failed component and is not a consequence of problems elsewhere. Fuel pump settings occasionally move and can result in one piston being severely overloaded compared to the others. The problem could also be in the installation, a faulty dynamometer or water supply, or a restriction in the inlet or exhaust. Birds' nests or even dead birds can cause a major obstruction in an exhaust or intake pipe and plastic bags are a real hazard in water systems.

It is hoped that the prototype engine does not seize or fail destructively and progress may be made with performance development or with identifying and perhaps resolving minor problems such as oil leaks. A worthwhile target is to achieve about 50 hours of running and then to dismantle the engine completely, even if progress with performance etc. is good. It is important to inspect all components early in the life of the engine becuase they may not be performing well, even if there is no immediate risk of failure. Bearings in particular can reveal much more information about alignment, oil supply, etc. when they have only run a short period of time, whereas with longer running these indications are combined into an overall pattern of wear. A particular instance is shown in Fig. 5 where rippling of the overlay bearing material is shown which probably indicates a marginally inadequate oil film thickness during some part of the cycle of operation. The damage is not catastrophic and with further running the damaged overlay material is worn off without necessarily resulting in failure of the bearing. By viewing the damage early, information about the oil film thickness is gained and consideration may be given to revising the bearing oil groove arrangment or the supply of oil to the bearing. There are now computer programs available which estimate the oil film thickness in journal and connecting rod bearings for alternative grooving arrangements and these are commonly used at the design stage. But all theoretical calculations involve assumptions which vary in accuracy in relation to different applications. Thus, the results of any theoretical analysis must be assessed alongside the practical results and it is not unknown for a theoretically less good bearing arrangement to perform best in practice. In solving such riddles it is most useful to mentally picture the component operation and fluid flow in slow motion so as to construct hypotheses which may guide a redesign, as this may provide a more convenient and rapid solution than rigorous theoretical analysis. The essential requirement is that sound decisions be taken early in the development as to whether a component is probably satisfactory whether it must be modified in some way in order to be adequately reliable. Redesigning, manufacture and inspection involve long lead times which can seriously delay a major project and which could prevent target completion dates being achieved. Within the commercial pressures upon most companies today, completion dates are extremely important.

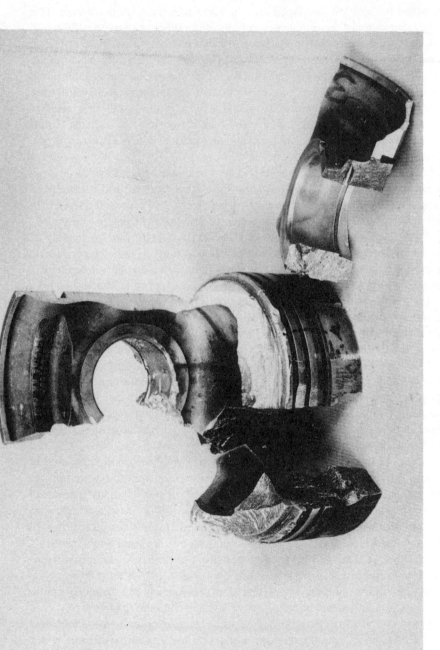

Fig. 4 – Chronic overload of a piston design has resulted in three simultaneous modes of failure. The crown of the piston has failed 'radially' by thermal fatigue; the skirt walls have cracked from the crown because of thermal arching of the crown; and mechanical loads have resulted in vertical splitting of the piston boss.

Fig. 5 – Rippling of bearing overlay material. The white spots show where the thickness and specification of the bearing metallurgy have been checked by the manufacturer. Rippling is probably caused by marginally adequate oil film thickness.

4. INSPECTION AIDS

There are certain techniques of measurement and aids to inspection which can greatly assist in deciding whether a component is likely to have adequate life and reliability. Temperature measurements are particularly useful on those components which form the combustion chamber of the engine, particularly the piston and valves, but there is immediately a problem because not only do these components have high surface temperatures but they also move and are surrounded by the engine structure. It is therefore exceedingly difficult to use thermocouples on these components, although some researchers have achieved it using linkages or radio telemetry. For general development, such sophisticated techniques are rather too expensive and inconvenient and so the information required is generally obtained either by hardness relaxation techniques or with fusible plugs. Hardness relaxation takes place in some steels and aluminium alloys when they are tempered and is dependent upon the temperature to which the metal has been exposed and the duration of exposure. This phenomenon may be used either by fitting a component with commercially available plugs of suitable steel which may be removed after test for hardness checking, or by measuring the hardness of the component itself (Fig. 6), provided it was made of a suitable alloy. Exhaust valves, for example, may be manufactured from hardened martensitic steel for a special test and often the standard aluminium alloy, as used for pistons, is suitable but then the component must be destroyed for the purposes of the hardness measurement. Alternatively, plugs of metals having different melting points may be sunk into a component and the temperature of an area estimated from a knowledge of which plugs have melted and which have not. Each of these techniques has its own advantages and the information gained can be most valuable. It is well established that the temperature of a piston at the top ring position should not exceed 230°C otherwise the lubrication of the ring may well fail, resulting in either scuffing of the ring or sticking of the ring in its groove. Some engines do run with higher temperatures than this but they usually employ specially coated piston rings of wedge section and these have other disadvantages. The limit for valve seat temperatures is usually 600°C for an engine burning distillate fuel or 550°C for a large engine burning residual fuel. Above these temperatures, carbonaceous or other products of combustion are deposited on the valve seat, preventing the valve from closing properly which in turn results in the combustion flame acting as a blow torch and burning the valve away.

Examination of the bedding patterns is an essential part of determining whether components are satisfactorily working together (Fig. 7). The primary question is whether any damage has occurred to the surfaces in the form of scuffing or scratching by debris. Dirt can be a major problem, but an engine must be capable of accepting small amounts of dirt without failing because it will almost inevitably occur on occasions in service. Debris can also result from the wear of some other component and so it is worth while determining the source of the

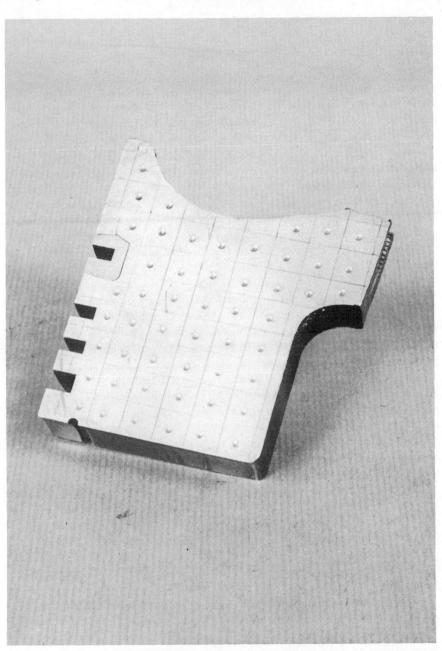

Fig. 6 – A section of a piston which has been hardness checked after engine test
as a means of determining temperatures through the section.

Fig. 7 – Graphite coating of the piston skirt reveals the bedding pattern very clearly. This prototype piston is close to failure from scuffing at the top of the skirt and over the intermediate ring lands. There is also some hollowness below the scuffed area and improvements to the skirt profile should be investigated.

debris by examination of the track it has left and the nature of the particles. Ferrous particles embedded in non-ferrous components may be identified chemically, which is most useful for bearings where the dirt may either be carried in by the oil or result from damage to that or some other bearing. Where the debris is a consequence of surface damage in the same component it is often important to know accurately the characteristics of the surface both in its new condition and after it has run. Modern surface measurement machines have been greatly enhanced by electronics so that a surface trace can be analysed automatically as it is taken and complex parameters calculated immediately. The usefulness of these new parameters varies with the application, but for important surfaces like cylinder liners, the traditional CLA value is no longer adequate. Obtaining a graphical trace of the surface does remain a most valuable technique, however, because a picture can always demonstrate complicated matters more simply than words. Using the trace in conjunction with experience of how the surface has performed in the engine should enable suitable parameters to be chosen and defined, but surface measurement is now a major subject which cannot be considered further in this chapter. Another recently available technique which may be of considerable assistance in understanding surfaces, and particularly their texture, is the scanning electron microscope. The three dimensional effect picture which it provides has caused many traditionally accepted machining techniques to be shown in a poor light and manufacturers are having to review their processes accordingly. The example shown in Fig. 8 illustrates how diamond honing does not cut cleanly into the surface of a cast-iron cylinder liner but rather ploughs and smears the cast-iron. Many small particles remain only lightly attached to the parent metal and these will be broken off early in the life of an engine, providing debris precisely where it can cause significant damage to the piston and piston rings. Since diamond honing has apparent advantages from a production point of view, it is a great asset to have a picture which so clearly shows the disadvantages, in a way which conventional optical and surface measurement techniques cannot, and so convince colleagues from other disciplines that change is necessary. A further method which is very simple but is most useful for chrome plated steel components is to dip the cleaned components into copper sulphate solution. If the chrome has been worn through in any area the bare steel will be lightly plated with copper but the chrome will not be touched, so that the degree of wear may be accurately assessed.

Modern design techniques using computers for involved calculations or sophisticated models to determine stress levels with good accuracy have considerably reduced the risk of major mechanical failure in components. Nevertheless it occasionally occurs, and it is vitally important that assurance is gained, that mechanical failure has not initiated. The first indications are usually in the form of a crack which may be small and inaccessible at first. It is therefore essential to determine first, where cracks are liable to occur and secondly, their extent, and to be able to distinguish cracks from scratches or other marks.

Fig. 8 – A diamond honed liner surface as viewed through a scanning electron microscope. The lack of adequate cross hatch in this instance indicates a machine fault but the 'foil' is typical of diamond honing.

Cracks are always a risk at sharp corners and other stress concentrations which are load carrying or are subjected to high temperatures. Thus, crankshaft fillet radii, connecting rod small and large ends, piston crowns and pin bosses, cylinder head injector holes, etc. should all be carefully examined by an experienced eye. If there is any suspicion of a crack then a non-destructive test is appropriate, and there are several to choose from. The most universal (but also the more messy) are the dye penetrant methods in which the suspect area is soaked with a highly coloured, free flowing dye. After a few minutes the excess dye is washed from the surface and a devloper used to draw the dye out of any cracks, which are then shown up as coloured lines. Casting porosity will also be highlighted by this method which can lead to confusion in some instances although this should be resolved by careful examination. A very useful and quick method of identifying cracks in magnetic irons and steels is to apply a magnetic ink (iron filings in suspension) to the suspect area and to introduce a magnetic field. Any magnetic discontinuity such as a crack or change of material is immediately highlighted because the iron filings move to lie along the discontinuity, identifying it as a clear line. This method works well using a reasonably strong permanent magnet so it can easily be applied in less than ideal conditions, although it is also used for production quality checks with greater sophistication. Eddy current methods of crack detection are similar to magnetic but are restricted to production applications because of the quality of equipment necessary.

Once a crack has been discovered the important work of deciding the reason for the failure begins. This will usually involve discussions with many colleagues leading to further measurements and investigations as decisions are made as to whether the material was to specification, the component made to drawing and the design satisfactory. Frequently with prototype components, the material and dimensions are not perfect because of the difficulties of manufacturing, but these frustrations must not be allowed to detract from the opportunity of reviewing the design. A truly reliable design will incorporate some reserve strength to allow for the inevitable variations which occur in production and service. In any component examination there is great value in endeavouring to understand the cause of every mark and imperfection which can be seen as this greatly increases confidence that reliability has been correctly assessed. Unfortunately such study takes a somewhat longer time than is usually available in today's environment, but discussions with more experienced colleagues will help to achieve the objective within an acceptable period.

5. DEMONSTRATING RELIABILITY

Despite the disappointments and frustrations of unexpected difficulties during initial prototype engine running, great efforts will be made to achieve a reasonably final engine specification at an early stage and to allow endurance testing to begin. It is important that endurance tests should begin as soon as basic engine reliability has been established and there is reasonable confidence, or at least hope, for the major components, because the established endurance test is the only method of ensuring that an engine will be satisfactory in service. Different manufacturers have developed their own cycles and unfortunately there are significant differences between these cycles. Since each manufacturer is relying on the experience gained from previous tests in order to make judgements, each insists on retaining the cycle with which it is familiar and which is known to have found faults with earlier designs. The concepts behind an endurance cycle are two-fold: to accelerate the wear and tear which the engine will experience in service by increasing the rating of the engine and to ensure that all components have undergone a sufficient number of fatigue cycles. Acceleration of wear and tear is usually achieved by increasing both the speed and the torque above that which the engine will produce in service, so as to increase the power by a minimum of 10%, and preferably more. Speed increases are particularly useful because the efficiency of diesel engines reduces significantly at speeds above the rated speed so that both thermal and mechanical loading are increased. Overfuelling is also important because of the significant numbers of customers who deliberately adjust the fuel pump to increase the power of an engine. The duration of the endurance test does vary, possibly beginning at 100 hours early in the development programme, whilst confidence is being established, and increasing to 500 or 1500 hours. One of the factors in deciding upon the length of the test is that many major components are subjected to cyclic stresses and therefore to the risk of fatigue failure. For the materials used in a diesel engine it is established that the cyclic stress which will cause failure decreases as the number of cycles increases up to 10^7 cycles but does not decrease for cycles beyond 10^7 (Fig. 9). Thus if any component is operated for more than 10^7 cycles and does not then show any indication of failure, it may be presumed to be safe for an infinite number of cycles. For an automotive diesel engine, 500 hours is usually long enough to have exceeded 10^7 mechanical cycles, but thermal cycles, from the engine being cold to reaching full temperature, are more difficult to accommodate. It is usually accepted that if the rate of change of temperature both in warming up and in cooling is made as rapid as possible, combined with the higher temperatures of overload running, then fewer than 10^7 cycles should be adequate to detect problems due to thermal fatigue.

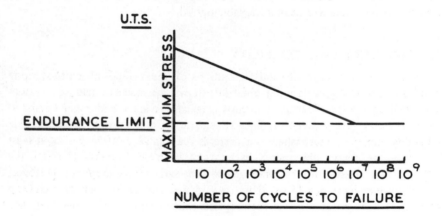

Fig. 9 – Derivation of a typical endurance limit which is reached at about 10^7 stress cycles.

In addition to all the above factors, the cycle chosen should recognize the varying demands upon the engine in service. Long periods of idling or running at low torque lead to greater amounts of oil being carried to the combustion chamber, which could in turn result in ring sticking or other unwanted deposits of carbon or lacquer. Thus, a good cycle (Fig. 10) will incorporate periods at each of the conditions which is known to cause some form of distress but will also arrange these in a sequence which provides thermal cycling. The overall objective is to accelerate the patterns of wear and difficulties found in service so

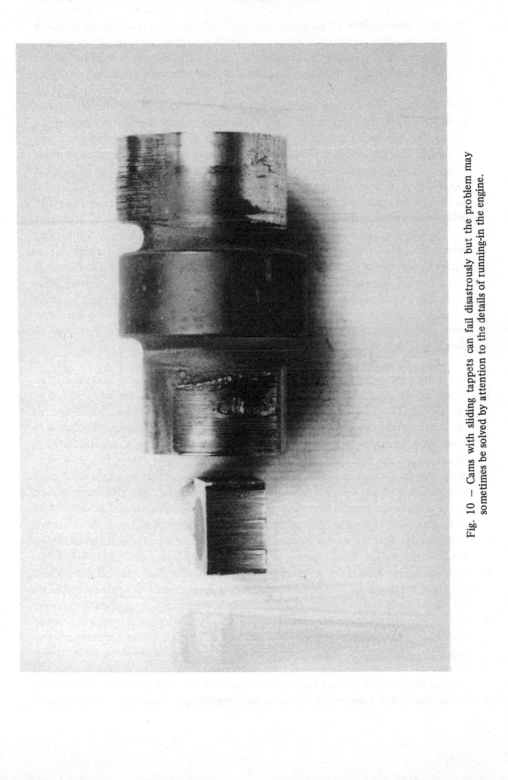

Fig. 10 – Cams with sliding tappets can fail disastrously but the problem may sometimes be solved by attention to the details of running-in the engine.

that they arise within an adequate time scale but not to so increase the demands on the engine that it ends up being over designed, and thus unnecessarily expensive. For this purpose, all wearing components must be carefully measured before the engine is built and the results recorded for comparison with measurements to be taken at the end of the test. Routine measurements to production standards are not adequate for this purpose because it is to be hoped that the changes in dimensions will be very small, so special arrangments should be made either within the Research Department or with the Standards Room. Confidence in the results is vital because metallurgical changes may cause a dimension to increase (Fig. 11), even though wear has obviously taken place. Judgement as to what degree of wear is acceptable can only be made in the light of experience but it should be remembered that the normally satisfactory wear pattern is for a high initial rate of wear during bedding in, followed by a stable period of low wear rate and finally a high wear rate at the end of the component life. Unsatisfactory wear may arise because the rate of wear never stabilizes at a low enough rate or because the stable period does not last long enough. Which of these factors is the cause of the problem will influence the choice of corrective action. Wear rates and initial clearances must also be assessed in the light of the tolerances which will apply in production, and care must be exercised to ensure that representative samples are tested so that the drawing tolerances can be shown to be acceptable.

The above review of the development of a new engine is only an outline of the numerous detailed investigations which need to be made as problems arise. Many of the areas of investigation require specialized knowledge and techniques which cannot all be mastered by a development engineer who usually spends much of his time in organizing a project and analysing the data and problems as they arise. There is therefore the need to determine as early as possible when a problem is such that guidance will be required from an expert in a particular area and also to know where the appropriate expertise is available. Larger companies may include a number of specialists within their structure, but on some occasions, more frequently for smaller companies, consultation will be necessary with representatives of supplier companies, academics or professional consultants. Thus arises the importance of good communication between those involved in engine development, as in many other facets of engineering. Communication is essentially an exchange of information so that all the parties involved are aware of the immediate and long term objectives of the project and all useful facts and information are made available to those who need to know. No expert or manager can be expected to contribute good advice or sound decisions if information is withheld from him or her but nor can technicians and testers be expected to work enthusiastically if the reasons for their work are not explained. The observations and comments of testers and even labourers should always be listened to and examined because their close involvement with the components and engines is a source of rich experience which must not be wasted.

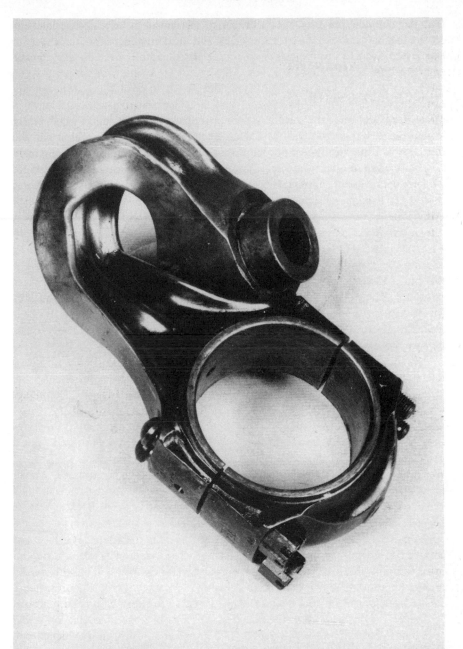

Fig. 11 — This engine has failed by hydraulic lock — possibly a failed gasket — so that the cylinder filled with water, which was effectively incompressible. As a consolation it demonstrates the ductility of the connecting rod material.

The development engineer is in the privileged position of gathering and co-ordinating information from all sources and initiating appropriate action or discussion. Only by ensuring good communications can he or she do the job well.

ACKNOWLEDGEMENT

Within that context, the author wishes to express his gratitude to his present colleagues at L. Gardner & Sons Ltd. and to many former colleagues at Mirrlees Blackstone Ltd., together with associates at component supplying companies, for the sharing of their knowledge and experience in the past and for assistance in the preparation of this chapter.

Index